An Archaeometallurgical Survey for Ancient Tin Mines and Smelting Sites in Spain and Portugal

Mid-Central Western Iberian Geographical Region
1990-1995

Craig Merideth

BAR International Series 714
1998

Published in 2019 by
BAR Publishing, Oxford

BAR International Series 714

An Archaeometallurgical Survey for Ancient Tin Mines and Smelting Sites in Spain and Portugal

ISBN 9780860548928 paperback
ISBN 9781407350301 e-book

DOI https://doi.org/10.30861/9780860548928

A catalogue record for this book is available from the British Library

This book is available at www.barpublishing.com

BAR Publishing is the trading name of British Archaeological Reports (Oxford) Ltd. British Archaeological Reports was first incorporated in 1974 to publish the BAR Series, International and British. In 1992 Hadrian Books Ltd became part of the BAR group. This volume was originally published by Archaeopress in conjunction with British Archaeological Reports (Oxford) Ltd / Hadrian Books Ltd, the Series principal publisher, in 1998. This present volume is published by BAR Publishing, 2019.

BAR
PUBLISHING

BAR titles are available from:

BAR Publishing
122 Banbury Rd, Oxford, OX2 7BP, UK
EMAIL info@barpublishing.com
PHONE +44 (0)1865 310431
FAX +44 (0)1865 316916
www.barpublishing.com

for "Z" zeta
1987 - 1993

who offered conviviality and elan

the years we spent visiting
and exploring

abandoned mines, mining towns
and archaeological sites

in the Iberian Peninsula

I would like to thank Prof. B. Rothenburg, Director of the Institute of Archaeometallurgical Studies (IAMS) for the opportunity to have been able to work with him over many years on archaeometallurgical excavations and post-excavation work.

Working with Professor Rothenburg was my primary introduction to the new and interesting field of archaeometallurgy The continued work with Prof. Rothenburg finally prompted me to take on the task of trying to organise and finish a PhD project.

I am grateful to Dr. J.F. Merkel, the supervisor of my PhD project, for all the time and effort that he has spent in helping me to carry through and finish my PhD project.

Dr Brinley Roberts of the Department of Geology, Birkbeck College was very generous with his time and advice in helping me to address and solve numerous geological problems arising from the survey.

I am grateful for all the help over the years from Dr. R.F. Tylecote who died during the course my PhD work, but whose lectures and discussions on archaeometallurgy had made strong and lasting impression.

I am grateful for the help in the form of lectures and personal discussions from Professor H. G. Bachmann, Professor J.A. Charles and Dr. Paul Craddock.

I would like to thank Sandra Bond, Deborah Glanfield and Kevin Reeves, Laboratory Technicians at the Institute of Archaeology, for all their help during the time I was doing my laboratory work.

Andy Beard, Microprobe Technician in the Geology Department, Birkbeck College, was most helpful in analysing very difficult mineral and metallurgical samples.

All survey field planning work at the Logrosan site in 1995 and 1996 was done by Brenda Craddock and Anthony Bridgeman. They were helped in 1996 by Jamie Thorburn.

Brenda Craddock has drawn all the final Logrosan site survey plans and all of my survey pottery and stone artifact drawings.

In Spain, I received invaluable help from Marcos A. Hunt Oritz and Victor Hurtado Perez of Seville University in helping to initiate preliminary contacts in Extremadura. I also received advice and help on geological problems from Felix Garcia Palomero of Rio Tinto Minera S.A., Rio Tinto, Spain.

In Extremadura, I received generous advice and was given invaluable geological publications by Carlos Alcalde Molero and Alfonso de las Llanderas Lopez of the Department of Direccion General de Industria Energia y Minas (Junta de Extremadura) Merida.

Tomas Porro Mayo, the custodian of the Museo Geologia de Extremadura, Merida helped to locate invaluable geological publications of Extremadura which he kindly allowed me to copy.

In Extremadura, those helping me with archaeological problems and advice were Alonso Rodriguez Diaz and Ignacio Pavon Soldevia (Universidad de Extremadura) and Jose Luis Mosquera (Archaeologo Regional Merdia). Augustin Velazquez Jimenez (Secretario) and Josefina Molina Garcia of the MNAR Museum Merdia were extremely helpful in locating finds from Logrosan and allowing me to have them drawn and sampled.

In Salamanca, I was helped generously by Dr. Angel Esparza Arroyo and Javier Larrazabal Galarza (Universidad de Salamanca) and Manuel Santonja Gomez (Museo de Salamanca).

In Portugal, Luis Oosterbeek (Escola Superior de Tecnologia, Tomar) was crucial in helping me acquire the preliminary permission for work in Portugal and in makeing initial contacts. Paulo Felix (Escola Superior de Tecnologia) helped with advice on pottery collected at Portuguese sites. Monge Soares Department, Quiminca Icenlineti and Fernando C.S. Real (Instituto Portuguese do Patrimonio Arquitectonico E Arqueologico) were helpful with advice and further contacts.

Raquel Vilaca (Instituto de Arqueologia, Coimbra) has been most helpful with archaeological advice and dating of survey pottery and allowing me to analyse her excavated metallurgical samples and to include the results in my PhD paper.

Maria Helena Simoes Frade (Diretoria Regional do IPPAR Coimbra, and Director of the Excavations at Torre Romano Centumcellas) has been most gracious in giving me her excavated slag samples from Torre Romana Centumcellas to analyse and allowing me to include the results in my PhD work.

I wish to thank the Central Research Fund, University College London, for providing money towards my preliminary survey trip in 1992 and the Institute of Archaeometallurgical Studies (IAMS) who also helped financially in supporting the first field survey trip in 1992 IAMS also later provided partial financial help in the survey planning trip to Logrosan during 1994.

I feel that I owe a debt of gratitude to the Extremadura Geologist, Vicente Sos Baynat, whose field work and writings have been invaluable to me in locating forgotten mineral and mine sites in Extremadura.

Vicente Sos Baynat died in 1994. I never had a chance to meet him, but his enthusiasm for work and his noting and saving of archaeological remains at mining sites, notably Logrosan, have been a major contribution to the development of my PhD work.

I also received invaluable help in the translation of my Spanish Permission Applications and annual survey reports from Marcos A. Hunt Ortiz, Seville and David Iguaz Hermoso de Mendoza, a Spanish PhD student at the Institute of Archaeology.

The reading and correcting of PhD thesis was under taken by Dr.J.F. Merkel and Stefan Lorett.

The subject of my still on going archaeometallurgical survey work includes my PhD thesis on the archaeometallurgical survey which I undertook in Mid-Central Western Iberia 1992-1995.

The area cover by the survey consisted of a geographical region including areas in both Spain and Portugal. The pre-historical period covered was from the Copper Period approximately 2000 BC to the late Roman Period the 4th Century AD.

The main theme of my work was and is to locate and assess surface evidence that the cassiterite ore in the Mid-Central region of Iberia had been mined, smelted and used in bronze production in antiquity.

The first task was to evaluate available historical and geological literature and to compile from the literature a list of cassiterite mining sites for actual field survey visits.

The primary season of field survey work consisted solely of trying to locate precisely the listed cassiterite sites and then to document if possible any archaeometallurgical surface evidence present at these sites.

After the first field survey season, I then undertook laboratory analytical examination of important collected samples of slags, metals and minerals using primarily SEM/EDS analysis.

The laboratory work was undertaken to provide positive element identification of the relevant collected metallurgical and mineral surface finds.

During 1993 and 1994, further survey field trips were completed and more surface material was collected from Logrosan, Golpejas, Valdeflorez, Area Catraia Do Buraco and Torre Romana Centumcellas for further laboratory analysis.

In 1994 and 1995, the predominant prehistoric cassiterite mining and production site at El Cerro de San Cristobal, Logrosan was planed on a scale of 1:500 in both surface and contour details.

During 1996 and 1997 the sites of Logrosan, El Cerro San Cristobal and the site of Mina de Berrocal were again visited for further surface study and planning work.

The Appendix II section of this paper was not included in my original PhD paper as not all of the analytical work on the metal samples had not yet been completed when the PhD was finished for presentation during the spring of 1996.

It is hoped in 1998 to start some exploratory excavation trench work and further planning on Cerro de San Cristobal and to start planning out a site plan of the Mina de Berrocal mine site.

CHAPTER 7

CHAPTER 8

The primary aim of this Archaeometallurgical Survey in the Mid-Iberian Peninsula was to document with surface evidence that tin mining and smelting took place in antiquity.

The starting objectives for this proposed survey were to visit and collect samples at modern listed tin mines or known cassiterite mineral areas current or ancient. The tin mining sites were initially identified either by published geological information on current modern tin mine locations or by earlier archaeological-metallurgical report surveys that had been undertaken within the Mid-Iberian peninsula region.

When I first started my survey project, I was not sure that the survey would actually produce any positive results and there existed the possibility that the actual field work might disprove what I was hoping it would prove. It was quite possible that the cassiterite mineral sites I planned to visit might end up offering no archaeometallurgical collection possibilities.

If surface finds were available at the sites that I was going to visit, I would collect samples of slag, furnace remains, pottery and some stone hammers, stone grinding tools and mineral samples.

I was not sure what amount of surface material that I would encounter. The amount of material to be collected would be determined only when I had actually arrived at each of the sites. In the end the amount of material collected did not really become a problem during my survey work due to the scarcity of material at the majority of the sites I visited. Therefore during the survey at many sites I collected all surface artefacts found, which in most cases were very few.

Only a few of the visited sites presented the problem of deciding what to collect. Archaeometallurgy remains occured in very small quantities, therefore all metallurgically related material was usually noted and collected during the survey.

The first year of field work in Iberia was done in 1992. The initial survey season took place during August and September. I chose to limit my survey work to within the Central and Southern areas of the Iberian Massif granite belt which runs from the North-West Spanish province of Galecia to the South Central provinces of Jaen, and Cordoba above the Gualadalquivir Valley.

Twenty seven sites were visited during the first survey season in the summer of 1992. In Spain the provinces of Badajoz, Caceres and Salamanca were visited and in Portugal the Provinces of Beira Alta, Beira Baixa and Upper Alentejo. None of the above six provinces have been totally surveyed.

Even today in 1995 with further library work, I am still finding new references to tin mineral areas which I did not know existed when I first started doing my field survey work.

As the survey work progressed over three years, I have with experience become more efficient at locating the actual mine sites.

Some of the published descriptions and locations of mines ancient and modern have not been very helpful with their sometimes inaccurate location details.

My first year of survey work in Iberia was a rather superficial exploratory nature due to funding limitations and travel expenses.

The primary objective during the first year's survey was to be able to physically locate the mine sites. The second objective was to document any evidence (no matter how diminutive) on the surface of the visited site.

The third objective was to identify dating evidence as well as metallurgical evidence to test the theory that prehistoric tin mining and smelting occured within the central Mid-Iberia geographical area. The next objective was to ascertain if my tin survey project would actually be worth the time and the effort for further fieldwork.

The first year survey in 1992 seemed quite promising, so I decided to continue with the project with further survey trips in 1993. In 1992 twenty seven sites were visited; some for only a few hours, others for one or two days. Surface evidence such as surface scatter of identifable pottery fragments had appeared at some sites and pointed to possible tin mining or smelting use during the Late Bronze Age and Roman Periods.

The second year of survey took place during 1993 and consisted of two trips to Iberia. The first trip was in March and the second trip was undertaken in June and July. Thirteen new sites in the Mid-Iberian area were visited during 1993. Also some of the earlier promising 1992 surveyed sites were revisited to study their geological settings in more detail and to spend more time looking for further surface evidence.

A third year of planning and survey was done in 1994. The work in 1994 again consisted of two separate trips to Iberia. The first trip was done in April-May and the second trip in December.

In the 1994 survey season the majority of the time spent in Iberia was at one site called Logrosan which was starting to look very promising as a possible Late Bronze Age tin mine and settlement site. I decided therefore to have a very detailed 1:500 scale plan drawn of this very large site; approximately eight hundred meters long by four hundred meters wide.

The site planning work at Logrosan was undertaken by Brenda Craddock and Anthony Bridgeman. The main instrument used for the planning work in 1994 was a standard (non-electronic) Hilger Watts ST 156 theodolite.

The detailed site plan of Logrosan was not finished in 1994 due to the complexity of the site in both geographical setting and the large number of surface features which had to be put on the plan. A further planning season at the Logrosan site was projected for 1995.

Only two new additional sites were surveyed in 1994; though once again some of the original surveyed sites from the 1992 and 1993 years were revisited to look for and collect possible surface information.

A four wheel drive vehicle was used in all the survey work. For a majority of the field sites a four wheel drive vehicle was not a necessity; except for making access to about 15% of the sites easier than it might normally have been.All other visited sites could have been easily accessible with any standard transport; provided that it was driven with care. All the actual site survey work that was done once the mine sites had been located was done exclusively on foot.

The majority of the actual site survey work was conducted unaccompanied. In Salamanca Province at the Golpejas Mine, I was helped by Javier Larrazabal Galarza an Archaeometallurgical PhD student from Salamanca University. He helped me with detailed field identification of pottery sherds.

I did gather some additional information on the visited mine sites from local people, if there were any present, when I was surveying the actual sites. The majority of the mine sites I visited and surveyed were devoid of any human activity during the majority of my visits.

As it turned out, the archaeometallurgical surface finds at 95% of the visited sites were of a very small quantity or even non-existent. Therefore all evidence that was found on the surface at any of the visited mine sites was collected; any interesting mineral samples, any stone tools, any shaped/rim pottery shreds and any possible smelting remains including slag, metal artefacts and furnace remains.

Most of the modern mine workings which I visited during my survey seem to have ceased production in and around the years 1977-1985. After 1985 it seems as though most if not all tin mines within the Mid-Iberian geographical/geological area had ceased production. It was during this period (1985) when the world price of tin dropped to its present day level.

Based on conversations with mine owners, it is doubtful; (even if the price of tin rises once again to the high levels of the early 1980's) that any of the Mid-Iberian tin mines will ever reopen. The percentage of remaining tin ore is low and higher wages levels now exist in Iberia. Tin ore minerals are mined and processed much cheaper in other regions of the world such as Bolivia, South East Asia and China (Lehmann B., 1990, 6-9).

As there should be no large scale modern destruction by mining of the ancient remains to take into consideration, future archaeological work including site planning and excavations at some of the mine sites which I have surveyed should be possible.

Many of the visited mine sites seem to have ceased work so long ago that local inhabitants did not have any first hand knowledge of past mine operations.

At some sites I visited, the local people did not have any knowledge of a mine ever being in their area or even what type of mineral had been mined in the past.

In 1995 I have started a fourth year of survey work. One trip was made during April-May. The main purpose of the first trip in April-May 1995 was to finish the plan of the Logrosan Late Bronze Age Tin mine/settlement site, which had been started, but not finished in 1994.

The planning work during the 1995 season was again done by Brenda Craddock and Anthony Bridgeman with help from Jamie Thorburn. In 1995 it was decided to use a Electromagnetic Distance Measurement (EDM) instrument. The instrument would speed up planning work on what was proving to be a very difficult site to plan.

The EDM used in 1995 was a Nikon Total Station DTM-20 with a 10/20 second digital readout. A Nikon DR-2 data collector was also taken along but not used. All survey points were recorded in hand written notebooks during the 1995 survey season.

The primary survey work on the Cerro de San Cristobal settlement site Logrosan was completed in the 1995 spring season.

The plan of the Logrosan site will probably never be satisfactorily detailed, due to the need for future excavation as well as the size of the site. A considerable amount of future survey work still needs to be done such as the precise planning of geological features including mine areas as well as ancient and modern galleries and shafts.

In 1995 further field visits and work were done on two other sites which I felt represented a good chance of being early tin mine/smelting sites: El Berrocal and Val de Flores. Both of these sites had been visited in previous survey seasons.

During the April-May trip in 1995 I started to use a small battery operated hand held satellite Global Positioning System personal Navigator for both finding given mine map co-ordinates and for fixing my actual survey positions more accurately. The small hand held unit I used was a Garmin GPS 40. The Garmin unit is able to work with both Longitude/Latitude or Universal Transverse Mercator (UTM) co-ordinates. Most current Spanish and Portuguese geological maps use UTM co-ordinates.

The accuracy of the GPS 40 location readings are between 15 meters - 100 meters depending upon the amount of degrading done by the American Military the satellite controllers and the extent of the daily satellite coverage in the area. The GPS 40 seems to have worked quite fast and accurately within the regions where I was surveying; due probably to the general open geographical nature of the landscape. I have now started to compile what is a more accurate location of Longitude/Latitude readings of my survey sites than I used in the first years of my survey work.

During all survey seasons all sites were photographed with 35mm colour negative and colour transparency film. Also some of the survey sites visited were also recorded on either 8mm or high resolution 8mm video tapes.

The colour negative film is useful for making paper colour copies that can be used with written report work. The transparencies will be used for future lectures. The videos which can show a more complete vista of sites have proved very useful in taking a second look at some of the sites when writing up my annual Spanish and Portuguese reports.

It is also hoped in the future to use the video tapes to make up a half hour to hour video report of my survey work. No black and white film was used during the survey expect in one or two instances; this was due to cost and the amount of time it was taking to do the colour photography and videos.

If any black and white photographs are required of the survey sites then I will go back to the selected sites at a later date. It is thought that the colour negative films will fulfill any functions that would have required black and white photographs.

The quality and scale of geographical and geological maps within the region which I worked are highly variable. I have not found any one set of maps either geological or geographical that would have suited all the requirements of my survey work. Therefore, I have had to obtain and use a variety of different maps.

For general travel in both Spain and Portugal I have found most useful Michelin Maps Carte Routiere Et Touristique series 1/400 000 No 441 North West Spain, No 444 Central Spain and No 437 Portugal.
The Michelin Maps have the major disadvantage of their large scale. However, the main advantage over the Spanish provincial maps is that they seem much more accurate in their correct geographical site placing of the symbols used for mine locations.

In Spain I have also used the Mapas Provinciales which are printed in the scale of 1:200.000. The provincial maps had to be used because the small scale Spanish maps 1:25.000 for most of the areas are still in the production phase and not available at present.

The provincial maps are very good in rendering the topography of the provinces understandable, but very poor in their mine location symbols. In some cases, symbols are marked on the maps kilometres away from their actual ground locations, or the symbols exist in areas where no mines seem to actually exist.

In Spain when available I have used the Mapa Topografico Nacional 1:25.000. These maps are the best for detail and helping to locate the modern mine sites. The problems with these maps is that with the scale being so small one must normally buy four maps to cover a certain mine area and its surrounding geographical region.

Due to the high cost of these maps it was only possible to buy maps for the mine areas that had shown some promise of being ancient mines or smelting sites.

A second problem with these maps is that quite a few of the 1:25.000 maps are not available in some areas or that the maps that are available are quite old with 1930's, 1940's, 1950's or 1960's publication dates.

In some cases it is the older publications which have proved more helpful in my survey work and finding mine locations as they show the areas during the periods when some of the mines were still working.

The older published 1:25.000 maps use Madrid as the Zero longitude and not the Prime Meridian at Greenwich England. This anomaly needs to be noted when giving map co-ordinates from the older published maps.

The ideal solution is to have both the old and new maps in the scale 1:25.000, but this can be a very expensive ideal option. If this option is not possible it then seems best from my experience to get the oldest dated published maps possible.

Also available in Spain are the Cartografia Militar De Espana Escala 1:50.000. These maps are up to date and of very good quality and carry both Latitude and Longitude and UTM co-ordinates. The using of the two co-ordinate systems on one map can make the map a little difficult to plot on at times. I have only bought and used these maps in areas where the 1:25 000 series maps are not available.

The Geological Maps for Spain are available at present in a scale of 1:50.000 and 1:200.000. The Spanish Geology service also provides economic geology maps in a scale of 1:200.000. Also available are single mineral economic geology maps for all of Spain on a scale of 1:500.000.

The 1:50.000 geological maps are excellent detailed maps but expensive and are not available for some of the mine areas which I surveyed. To date I have obtained six of these maps and they have proved very helpful in acquiring an understanding the geology of certain mine areas.

The 1:200.000 geological maps do not have the detail of the small scale maps, but have proved very helpful in understanding the larger regional geology of the areas in which I am working. Again there are certain areas for which these maps are not available as they are still in production. The 1:200.000 Metalgenetico Maps are excellent in providing good and fairly exact details of all mineral/ore deposits within the regions I was surveying.

I found very helpful for my preliminary survey work the Geology Map Cartagrafia Geologia Del Hesperico Del Sudeste De La Peninsula Iberica - 1:500.000 - published in 1975 (El Departmento De Petrologia Y Geoquimica De La Universidad De Salamanca). It offered an extremely excellent overview of most of the area that I surveyed.

The geographical maps used in Portugal were provincial maps 1:200.000 in scale. These maps are quite good, but the large scale hindered their usefulness in the field. More advantageous in the field survey was the 1:100.000 maps in scale as they contained more area detail, which was useful in helping to locate mine sites.

The main maps that I used, once an area had proven to contain a possible ancient mine, were the Portuguese Military maps of a scale of 1:25.000. These maps like the Spanish Military maps contain Longitude and latitude and UTM co-ordinates. The detail on these 1:25.000 Military maps is excellent.

The geological maps used in my survey work in Portugal were quite sufficient for archaeometallurgical survey use.

A very large scale 1:1.000.000 geology map covers the whole of Portugal. This map was only useful in giving a very good general geological overview of Portugal.

More useful were the Carta Geologica de Portugal maps in a scale of 1:50.000. These smaller scale geology maps are of a very adequate quality. The 1:50.000 geology maps all are accompanied by a small pamphlet written in Portuguese which offers general information on the geological genesis and mineral ores contained in the area that each map covers.

Again because of the vast number of geology maps that would have been needed to cover the complete area I was surveying; the 1:50.000 maps could only be obtained for areas that had offered proof of ancient mining or smelting activities.

For ease of future reference, the Spanish Geology Maps can be obtained at:

Instituto Tecnologico Geominero De Espana
c/ Cristobal Bordiu, 34
28003 Madrid
Tel: (91) 349 57 30
Fax: (91) 349 57 62

Spanish Geographical Maps can be obtained in offices of the Centro Nacional de Information Geografica to be found in most large Spanish Provinical Town capitals and Madrid.

Spanish Military Maps can be obtained from Militar Comisiones Geograficas located in Sevilla, Valencia, Barcelona, Burgos, Valladolid, La Coruna, and Granada.

Portuguese Geological Maps can be obtained at:

Servicos Geologicos De Portugal
R. Academia Das Ciencias 19-2
1200 Lisboa
Tel: (01) 36 39 15

Portuguese Geographical Maps can be optioned in most large Portuguese provincial towns as can the Portuguese Military Maps.

In this chapter I will be listing in chronological order and writing brief descriptions of the main books both historical and current writings which I have used for identifying sites for my survey visits.

The first written works which mention ancient mines and use of minerals in Iberia are the work of the Romans: Pliny, Diodorus Siculus and Strabo. Their descriptions are brief but speak of a land rich in mineral wealth with mines of lead, iron, copper, silver, gold, tin and cinnabar (Pliny, 25; Diodorus 191-203; Strabo, 39-51).

The following directly quoted passage by Diodorus Siculus is very emphatic in its description of the amount and placement of the tin mineral cassiterite present in Roman Iberia.

"Tin also occurs in many regions of Iberia, not found, however, on the surface of the earth as certain writers continually repeat in their histories, but dug out of the ground and smelted in the same manner as silver and gold" (Diodorus Siculus, 201).

The earliest modern book of an archaeometallurgical survey in Iberia is the pioneering work of Oliver Davies, Roman Mines In Europe. Davies was a lecturer in ancient history and archaeology at Queens University, Belfast, Northern Ireland. Davies book was published in 1935.

Most of Davies' field and research work for the book was done in the late 1920's and early 1930's. The main theme of Davies survey field work and publication was to present all the known Roman mines and mineral areas that had been worked within the boundaries of the Roman Empire at its greatest extent in the second century AD.

Davies survey book covers Italy, Gaul, Spain, Portugal, the British Isles, Rhineland and Upper Danube Provinces, the Illyrian Provinces, Dacia, Moesia, Macedonia, and Thrace and Greece. The book is no longer in print, though it is still held in some libraries.

Davies writings on tin in Iberia consist of only a brief mention consisting of a few sentences and a second small section consisting of only a few pages (Davies, 1935, 95, 103-5). Davies quotes Diodorous Siculus' writings about hard rock tin mining taking place in Iberia and the fact that tin was being traded from Tarshish, (which could most probably be equated to pre-historic Tartessus), but also states that no one seems to know as of yet exactly from which area the tin was being mined (Davies, 1935, 104).

Between 1935 and the late 1970's it seems as though very little further archaeometallurgical field work was done in the Iberian peninsula probably due to the general political chaos caused before and during the Spanish Civil War (1936-1939) and the general economic decline which took place in Iberia after the Civil War and during the Second World War (1939-1945). Spain did not really start to recover economically from the effects of the Civil War until the mid-1950's.

The next archaeometallurgy survey work, after Oliver Davies, was done between 1974-1976 by Beno Rothenburg (Institute for Archaeo-metallurgical Studies; University College London) and Antonio Blanco-Freijeiro (Madrid), Ancient Mining and Metallurgy in South-West Spain published (1981).

This very informative archaeometallurgy survey is of a more localised nature than Davies' 1935 book and covers only the Spanish province of Huelva. The metals which this work examines are copper, silver, iron and gold. The tin mineral cassiterite does not appear geologically within the province of Huelva.

In 1986 R.D. Penhallurick, an assistant Curator at the Royal Institution of Cornwall's County Museum in the town of Truro, published Tin In Antiquity. This book published by the Institute of Metals London concerns the mining and trading of tin throughout the ancient world covering Africa, Asia, Europe, and South/West England (Cornwall, Devon and the Isles of Scilly).

The Penhallurick book contains a section on Tin in Iberia (95-105) which covers both Spain and Portugal. The book proved most helpful in deciding which tin sites to visit in my first year's survey route, though the very vast scope of the book meant that any information provided on individual tin sites in Iberia was very brief and had to be supplemented with information from other publications.

The Penhallurick book also proved most useful in providing information and locations on other ancient (Bronze Age) tin production areas within Europe and the Mediterranean regions.

In 1987 the most thorough and comprehensive archaeometallurgical survey book to date on Iberia was published in French by Claude Domergue, Catalogue des mines et des fonderies antiques de la Peninsule Iberique Tome I & Tome II. This very substantial work in two volumes of five hundred and ninety seven pages covers over five hundred and seventy-five mining and smelting sites in both Portugal and Spain.

The Domergue archaeometallurgical survey covers the Copper Age, through to the end of the Roman period. Domergue states that he had visited ninety percent of the sites during his survey field work. The minerals included in the survey are gold, silver, copper, lead, tin, iron, mercury/cinnabar and zinc.

The majority of the mine/smelting mineral sites covered in the Domergue survey are silver, lead, gold and copper mining or smelting.

Of the total number of sites surveyed by Domerque only seventeen relate to tin mining and smelting. The total number of tin mine/smelting mineral sites actually visited and surveyed in Spain is ten sites; seven of the ten sites are located in North West Spain within the provinces of La Coruna (4), Pontevedra (1), and Orense (2).

The other three tin mine/smelting sites are located in the provinces of Murcia (1), Badajoz (1) and Salamanca (1). There are a further seven tin/smelting sites Domergue surveyed in Portugal; in the northern provinces of Minho (4), Tras-O-Montes (1),and in the mid eastern provinces of Beira Alta (1),and Beira Baxia (1).

In 1987 there also appeared a publication by J.C. Edmondson in English Two Industries in Roman Lusitania-Mining and Garum Production. This B.A.R. publication I believe is based totally on library and museum investigations. This very interesting work covers uniquely the Roman Iberian South Western province of Lusitania which includes about two thirds of modern day Portugal from the Duero river southwards and parts of modern day Spain provinces of Badajoz, Caceres and Salamanca and part of Zamora.

The publication consists of research into two very different subjects; the first subject being the infrastructure and production techniques both social and industrial of the Roman Lusitanian mineral mines.

The second subject of the Edmondson's publication is the production for export of Garum. Garum is a rich fish sauce made from small fish and was produced mainly in the Roman provinces of Greece, North Africa and Iberia and was then traded in ceramic amphorae throughout the Roman Empire (Whitehouse, 1983, 182).

Edmondson examined Roman mining and metal production of copper, lead, silver, gold, iron, manganese, and tin among others. Edmondson's writing is interesting in that, besides the technical aspects of mining in the Roman period, he also included the social and political aspects of the Roman mining areas within Lusitania.

Edmondson was able to present this aspect of Roman mining life due to the thorough and very informative archaeometallurgical excavations which have taken place at Aljustrel, located in the Southern part of Portugal within the South Iberian copper belt. Edmondson's publication lists a total of one hundred and twenty six Roman mining sites. Of this total, twenty-three are listed as tin mining sites.

During my survey work I found one or two sites (within visited areas) mentioned by Edmondson very difficult to find precisely. This could be put down to two reasons. Either Edmondson's literary sources are so old that the actual surface evidence mentioned has since disappeared or that I have visited a different area than that which Edmondson had written about.

In 1988 J. de Alarcao Professor of Archaeology at the Institute of Archaeology, Coimbra University published in English a set of two volumes in four books called Roman Portugal. This work by Alarcao is a very comprehensive survey of all known Roman sites within the borders of the modern state of Portugal.

The Alarcao publication consists of a Gazetteer of over three thousand Roman sites with a short description of each and eight accompanying maps. The survey covers mostly Roman towns and villas, but also includes Roman mines and smelting sites of copper, lead, silver, gold, tin and iron.

In 1990 Claude Domergue published, in French, a second mine survey book, Les Mines De La Peninsule Iberique Dans L'Antique Romaine. This very substantial publication (six hundred and twenty three pages) deals chiefly with the Roman period within the Iberian Peninsula. Domergue writes on prospecting, mining (both open cast and gallery systems), mineral preparation, smelting activities and administration of the mining areas.

The minerals/metals examined are gold, silver, lead, copper, tin, iron and mercury. A lot of the information in the Domergue 1990 mine survey publication had also appeared in his 1987 publication, which I mentioned earlier. The 1990 book also contains a vast amount of new information covering the prospecting and technical side of Roman mining activities within Iberia during the Romans' six century rule of Iberia.

Since 1976 considerable political power has been devolved from Madrid to the autonomous communities of Spain. Since 1976 the autonomous region of Extremadura, which is composed of the provinces of Badajoz and Caceres has produced three publications on minerals and mines located within Extremadura.

The main justification for the publications was to try and get commercial mining companies, Spanish or European interested in exploiting the mineral wealth of Extremadura and therefore creating employment in one of Spain's poorest regions.

The first publication in this series is in Spanish, by the Direccion General de Industria, Energia y Minas of the Junta De Extremadura, Marzo 1987 and titled La Minera En Extremadura. This publication included a second volume called Mapa Geologico-Minero De Extremadura.

The second volume was accompanied by two maps both on a scale of 1:300.000 titled I Minerales Metalicos y Energeticos, and II Minerales y Rocas Industriales. This very useful publication is now out of print and has been replaced by the May 1990 publication listed below.

The main volume of La Minera En Extremadura has a small section covering the history of mineral exploitation and historical mines that are known with in Extremadura (pp 12-18). The publication covers all economic metal/minerals (Pb,Zn,Cu,Au,Ag,Hg,Sn,W,Mo,Bi,Cr,Fe, Mn,Sb,P,U,Ba) and coal (Hulla-Antracita) found in Extremadura and any known working and non working mines within Extremadura.

The listings and map locations of the mineral and mines of tin in the two volumes of La Minera En Extremadura and accompanying maps has proved to be invaluable for my survey work.

A second edition of La Minera En Extremadura was published in Spanish by El Consejero de Industria y Turismo; Junta de Extremadura in May 1990. This edition consisted only of a single volume and two maps on a scale of 1:800.000 bound within a single volume.

The 1990 edition contains an interesting section listing all the published PhD research papers concerning the geological formation and minerals in Extremadura from 1941-1992. The historical section remains almost a duplicate of the 1987 edition. From a archaeolmetallurgists viewpoint the 1987 publication with its accompanying maps is a more useful publication.

Minerales Gemologicos De Exremadura (1993) by the Consejeria de Industria y Turismo; Junta de Extremadura is the third publication. This single volume work is accompanied by a Mapa Gemologicos De Extremadura in a scale 1:400.000 and covers all of Extremadura.

This book is on a smaller scale than the 1987 and 1993 publications. It lists and describes some mines that I had visited during my survey. There was some new useful information from this geological survey which had not appeared in any other publications.

All three of the geological-mineral publications on Extremadura were given to me by the Direccion General de Industria, Energia y Minas, Merida during my PhD work visits in Extremadura.

A second edition of La Minera En Extremadura was published in Spanish by El Consejero de Industria y Turismo; Junta de Extremadura in May 1990. This edition consisted only of a single volume and two maps on a scale of 1:800.000 bound within a single volume.

The book No 2 Estano also contains a description of the geological formation of the major listed mine sites. This book has proved very helpful locating tin mine production areas within Salamanca province. My survey work did not include any visits to Zamora, Segovia or Avila provinces.

In 1992 the Direccao-Geral de Geologia e Minas-Servico De Fomento Mineiro of Portugal published a pamphlet called Portugal Exploration And Mining; An Interesting Challenge examining aspects of administration and infrastructure, compiled by Lourenco, Goinhas, Martins, Viegas, and Gonzalez. This interesting but brief pamphlet (29 pages) lists all the minerals and geological location of the minerals within the boundaries of modern day Portugal. It also shows the location of all major modern mines going back to the 19th century, both working and non working. The section on tin was very helpful for planing my survey routes.

The aim of this chapter will be to offer to the reader a very brief background of tin in antiquity and its uses.

Tin was one of the important metals man has used in both pre-historical and historical times. The other main important metals used in prehistory were; gold, silver, copper, lead, and iron.

Metallic tin was normally alloyed with copper to form bronze and some what later in certain areas during antiquity lead was also added to the tin copper mixtures to allow better flowing capabilities for casting objects (Scott, 1991, 26).

It is probable that the first metallic minerals used by early Man were all native metals; native gold, native silver and native copper. Native metals did not require any metallurgical processes to make them useful, other than possible annealing and cold hammering. All most all of the above surface rich native minerals have by now disappeared due to two factors: their easily visible geological placement and their very pure mineral content.

The second most probable step in metallurgical development after the use of native minerals was the smelting of oxide copper ores and lead sulphide ores and further refining and working or casting of the purified copper. Lead was probably the first metal which man used and altered by a metallurgical reduction process.

In an archaeological context, copper is first dated to around 7000 BC - 6000 BC. It is probable that only native copper was used to produce most of these rare early copper artefacts that have appeared in stratified archaeological excavations.

By 3500 BC copper started appearing regularly within archaeological excavated sites in the ancient near East (Tylecote, 1976, 5).

All of the early copper artefacts would certainly have come from either native copper or copper carbonate ores, malachite ($CU_2[(OH)_2|CO_3]$ and azurite ($Cu_3(CO_3)_2(OH)_2$ being the most probable first copper carbonate ores used by early man.

Pure copper worked on its own either as a native or smelted metal can produce a hard edge, if the right amount of working and annealing is applied.

The intermediate stage between man's use of copper and bronze was the use of arsenical copper alloys. The main question still open to discussion on arsenical copper is "was the arsenic being mined separately or was the arsenic coming from copper ore deposits which had a varying percentage of arsenical mineral already present in the copper deposit" (Tylecote, 1992, 10). It is probable that both of the above conjectures apply to the early metallurgical use of arsenic.

Tylecote 1992 tended to believe that the arsenic mineral was present with the copper ore by natural secondary enrichment of copper deposits and not sought or mined separately by early man.

The amount of arsenic found in copper artefacts can vary from 2% to 7%. Artefacts with an arsenic content below 2% could be considered to have occurred as a result of a natural impurity in the copper ore used and artefacts with a arsenical content above 2% could be considered as a arsenic ore addition added to help benefit the hardness of the final copper product (Tylecote, 1992, 18, 26).

The use of arsenic with copper from analytical evidence seems to have been tied to the use of oxide copper ores for smelting with charcoal under reducing conditions.

Little of the volatile arsenic mineral up to a 7% limit would be lost in the smelting and casting of oxide ores containing arsenic. If the smelted copper/arsenic cast objects were then hot worked to induce hardness the arsenic and the hardness effect it brought to the copper would be lost (Tylecote, 1992, 10).

The same cannot be said of copper sulphide ores which contain arsenic, as they would have to first be roasted to lower the sulphide content which would have presented problems of arsenic retention in the copper during the smelting process (Penhallurick, 1986, 4).

The volatility of arsenic plus the health hazard presented by the poisonous arsenic fumes during smelting and the shift to sulphitic ores all probably helped to generate the search for an alternative hardening element, which could be used in copper smelting in place of arsenic.

Arsenical copper with the percentage of arsenic in the region of 5% - 7% will cast almost to the same hardness as pure cast copper. However as stated above a cold worked arsenical-copper mixture will make a much harder and usable edge than a cast worked pure copper tool. The hardening effect of arsenic on copper will happen quicker during the working process than when hardening pure copper only. It was more preferable to work the arsenical-copper cold as the percentage of arsenic retained within the worked object would remain constant (Tylecote, 1992, 10).

"The optimum temperature for hot working of arsenical copper was 300^{0}C; if the temperature during hot working goes higher any arsenic in a oxide arrangement within the copper will became very volatile and start to decrease in concentrations at the surface" (Clayton, 1974, 79).

Both arsenic and tin have a de-oxidising effect on copper and besides making it a harder metal also allow for it to be easily cast and easily hammered into a usable hardness.

How early man discovered that the arsenical-copper mixture was superior to pure worked copper and then later discovered that the addition of tin to copper could make an even a more durable and usable metal than pure worked copper are still not satisfactorily known.

There also appears in the archaeological circumstances during the early part of the copper/tin period copper/tin artefacts with traces of arsenic. The traces of the arsenic are not high, only 1%-3%.

There are two possibilities why this is happening. Firstly it is probable that copper-arsenical artefacts were being broken up and remelted and recast into copper/tin bronze artefacts with the further alloying addition of tin. The traces of arsenic present in the finished bronze artefact could be what is left over from the earlier copper-arsenic artefacts.

The second possibility is that copper/arsenic ores were still being mined and smelted, but tin was being added during the casting process to make true tin/bronze. An assumption could also be made that early man did not really understand the mineralogy of arsenical-copper ores, but thought the arsenical/ore mineral he encountered to be a true copper ore.

The technological achievement of alloying tin oxide (SnO_2) with copper either as an ore or in a smelted pure metallic form to make bronze was one of the major development steps made by prehistoric man in his use of metals.

The smelting of high purity cassiterite with charcoal can require temperatures as high as 1000^0C. The smelting of high purity cassiterite would produce a very pure tin ingot without producing much slag in the process. If any tin slag had been produced in antiquity it is also probable that it would have been subsequently finely crushed and re-smelted to recover any metallic tin prills still present (Tylecote, 1987, 141).

Pure tin has a melting point of 232^0C and when mixed with copper the melting/casting temperature of the copper-tin/bronze would be in the region of 950^0C - 980^0C.

Bronze, the mixture of copper and tin, started appearing regularly around 3000 BC within the archaeology context of the Middle East, but large numbers of bronze tools and artefacts did not become wide spread in archaeological excavations in the Middle East untill about 2000 BC (Ben-Tor, 1992, 206).

The gradual introduction of bronze took place at varying dates in different parts of the ancient world (Tylecote, 1976, 14-15). As with all things that have appeared in archaeological excavations; absolute dates are always not so "absolute" and one must consider the gradual progression that normally took place in prehistory. The gradual progression was influenced by either one or all of the following factors: economic trade, culture exchange and war.

A fundamental question in the study of ancient tin metallurgy is whether the tin oxide cassiterite was smelted with Cu minerals minerals or whether the cassiterite ore first was smelted on its own with charcoal to produce a pure tin metal ingot. The tin ingot would be alloyed with pure copper, most probably in a crucible to cast Bronze objects.

Both of the above methods were probably used to produce bronze in antiquity. It is probable that the pure cassiterite mineral was smelted directly with copper ore or metal near the tin mine areas and at the same time that the tin mineral in a separate process was also being smelted into tin ingots to be used for trading purposes.

Evidence of tin smelting from antiquity are very rare to date. There have only been two instances of tin slags being found within an archaeological context or excavated area. Slag is normally a waste product of the process of smelting mineral ores into metal; slag is a by-product and collector of the impurities fuel ash and gangue which occur when smelting mineral ores to metal. Usually a benefaction crushing/sorting process of the ore was used to remove most of the excess gangue before smelting (Bachmann, 1982, 9).

Ancient slag were formed from fluxes which were added to the smelting process to help the gangue material to chemically and physically separate more efficiently from the mineral ore which was being smelted. The main fluxes used in antiquity were iron and manganese oxides, lime, silica and perhaps fluorspar (Bachmann, 1982, 10).

Another point is that any native mineral or very rich ore used in antiquity gold, silver, copper, iron, lead or tin when smelted will produce almost no slag (Tylecote, 1987, 307). Properly sorted cassiterite ores from alluvial, elluvial or vein workings in antiquity would have been very rich, (probably somewhere in the region of 70% pure). Therefore if and when cassiterite was smelted with charcoal in antiquity, it would most likely have left little or no trace of slag.

Cassiterite ore in some instances can be found very pure or is easily sorted to a pure state by crushing and further hand or wind sorting. This pure tin ore could then have then been added directly to an already smelt/refined copper in a crucible smelting process to produce a good casting tin/bronze (Franklin, et al, 1978, 29).

Of the two ancient tin slags recorded, the first slag comes from what is thought to be a Early Bronze Age ritual burial (dated by pottery to 1500 BC) at Caerloggas, near St Austell in Cornwall, South-West England. According to Tylecote, the Cornwall slag was composed of a glass phase with small prills of metallic tin measuring approximately 0.65 um in diameter. Tylecote did not mention the amount of slag found (Tylecote, 1987, 307-9).

The tin content of the Cornwall slag is 40.4% accompanied by 4% iron oxide, 31.6% silica, 13.1% aluminium oxide, 1.5% titanium oxide, 0.7% magnesium oxide, 0.2% phosphorus pentoxide and 3.0% tungsten oxide (see Figure 3.1).

The second sample of tin slag comes from an excavated hill fort (dated between the second to eighth century AD) at Carvalhelhos in Northern Portugal (about 23 kilometres to the South-West of the town of Chaves). The excavators believe that the slag is from the second century AD Roman levels of the site.

The site of Carvalhelhos produced tin slag classified in two different sizes: the first being finely crushed fragments with a total weight of 1570 kg; the second size being pieces of slag weighing 0.3 kg - 0.4 kg each. It is not known how much total weight or pieces of the second size of tin slag was found (Beagrie, 1985, 167). The composition of the Carvalhelhos slag is as follows 1.8% Sn; 21.5% SiO_2; 9.3 $Al2O_3$; 57.5% FeO; 10.4% MnO_2; 0.96% TiO_2; 0.3% PbO (Costa and Santos 1965-1966).

%	Caerloggas (Cornwall) 1600 BC*	Cavalhelhos (Portugal) 8th cent. BC-2nd cent. AD Iron Age (Costa and Santos 1965–66)	Retallack (Cornwall) Medieval?	Hexworthy (Devon) Medieval?	Calenick (Cornwall) 18th–19th cent. (Tylecote 1976)	Stable Hobba (Cornwall)	Mod. tin slag (Bray 1947: 445)
SnO_2	(SnO)40.37	(Sn) 1.6	26.51	(SnO)10.15	(Sn) 7.9	22.0	(Sn) 2.5
FeO	3.97	—	9.35	10.33	32.6	27.79	18.0
SiO_2	31.56	21.46	40.25	48.61	30.0	31.29	35.0
Al_2O_3	13.15	9.34	13.66	17.13	11.1	10.26	16.5
MnO_2	—	(MnO)10.40	0.69	—	(MnO)0.2	0.37	—
TiO_2	1.54	0.96	1.42	3.50	—	0.97	—
MgO	0.74	—	1.07	—	—	0.50	17.0
P_2O_5	0.23	0.30	—	1.53	0.05	0.76	—
WO_3	3.08	—	—	—	—	5.55	—
CaO	—	—	4.24	0.50	1.6	0.87	28.0
K_2O	—	—	0.09	2.03	—	0.73	—
Na_2O	—	—	—	—	—	—	—
S	—	0.005	—	—	—	—	—
Fe_2O_3	—	64.43	—	—	3.9	—	—
Cu	—	—	—	—	1.48	—	—
Pb	—	—	—	—	0.02	—	—
Sb	—	—	—	—	<0.01	—	—
As	—	—	—	—	<0.01	—	—

* This slag was analysed by Dr C. Salter of Oxford University and I am grateful to him for allowing me to use it prior to his publication. Those from Retallack, Hexworthy and Stable Hobba were analysed by Miss E. Photos to whom I am also grateful.
— = *not sought.*

Fig 3.1 Composition of Tin Slags, Tylecote, R.F., 1987, 309

Tylecote (1987) thought that the tin slag from Carvalhelhos with its very low tin content, was a very advanced tin slag for its dated period; though if its dated period is the eighth century AD and not the second century AD then it would fit its element analysis easier.

Tylecote (ibid) felt that the Cavalhelhos tin slag in its analysis resembled very closely the analysed tin slags from the excavated remains of the Medieval tin blowing houses in Cornwall (Tylecote, 1987, 308-10). Some of these Medieval slag analyses are included in Tylecote's analytical table along with the two tin slags from antiquity.

A second possible explanation exists for the low tin content of the Carvahelhos tin slag. The slag may have been produced by iron smelting and the iron ore contained some tin minerals and the slag was later being crushed to remove the tin content (Tylecote personal letter to Maria e Costa, 1988).

"When adding sorted high grade cassiterite ores directly to molten copper, it would have been necessary to first cover the molten copper with charcoal and then add the cassiterite. This would allow the molten copper to accept more than the normal 1% of cassiterite (SnO_2)" (Cleuziou and Berthoudt, 1982, 15-16). The amount of cassiterite being added most probably was determined by weight or volume. The cassiterite ore would then be covered with a second layer of charcoal .

This technique of making bronze would mean that the cassiterite ore was being first reduced before it would flow down as molten tin metal into the lower part of the crucible and mix with the pure molten copper to make bronze.

Alternatively to make tin ingots cassiterite could have been smelted in a furnace with charcoal. The recovered tin metal would be refined and cast as pure tin ingots. This process would have taken place near the areas where the cassiterite had been mined. It would have been easier to transport the smelted tin ingots than the unsmelted mineral ore over any great distance. It is possible that rich refined unsmelted cassiterite ores were also traded, but probably over shorter distances than refined metallic cast tin ingots.

In the Late Bronze Age it is probable that the major trade of tin was done using tin ingots. The trade of tin in cast ingot form in antiquity is backed up by the remains of tin ingots along with a large amount of copper oxhide ingots in the Late Bronze Age underwater ship wreck off the south coats of Turkey at Ulu Burun (1200 BC - 1300 BC) (Bass, 1987, 693-733).

The occurrence of pure tin objects and ingots in excavated archaeological contexts is very rare. The following list includes examples but not all of the unalloyed tin ingot finds from around the Mediterranean and Europe.

a) bangle made from two twisted strands of thin wire; [found at Thermi - Island of Lesbos, Greece:cira 1800 BC] (Franklin, et al., 1978, 25).

b) a cast tin bottle (with six per cent lead present) and a tin ring; [found in a grave at Abydos, Egypt: cira 1500 BC] (Franklin, et al., 1978, 26).

c) two tin ingots inscribed, with Cypro-Minoan script; [found off the coast of Northern Israel: circa 1500 BC-1100 BC] (Franklin, et al., 1978, 45).

d) clay funerary vases from the Aegean; externally covered in tin; presumably as a substitute for silver; [Minoan and Mycenean: circa 1400 BC - 1350 BC] (Franklin, et al., 1978, 26).

e) a mug, a pilgrim flask and a bent plate and round bun ingot and four tin ingots; shipwreck [Ulu Burun, Coast of Southwest Turkey; cira 1300 BC] (Bass, 1987, 702-13).

f) possible badly corroded remains of a tin ingot (?) shipwreck [Cape Gelidonya, Coast of Southwest Turkey: cira 1200 BC]. (Franklin, et al., 1978, 26)

g) tin beads found at Sutton Veney in Wiltshire England [end of the English Bronze Age: cira 800 BC] (Franklin, et al., 1978, 26).

h) tin beads in an amber and faience necklace; Odoorn Necklace [found at Drenthe, Holland: circa Bronze Age] (Franklin, et al., 1978, 26).

i) a total of thirty two thin round shaped tin ingots; the ingots weighed approximately forty six grams each; [a shipwreck off the southern coast of France near Narbonne; cira 800 BC (?)] (Penhallurick, 1986, 106).

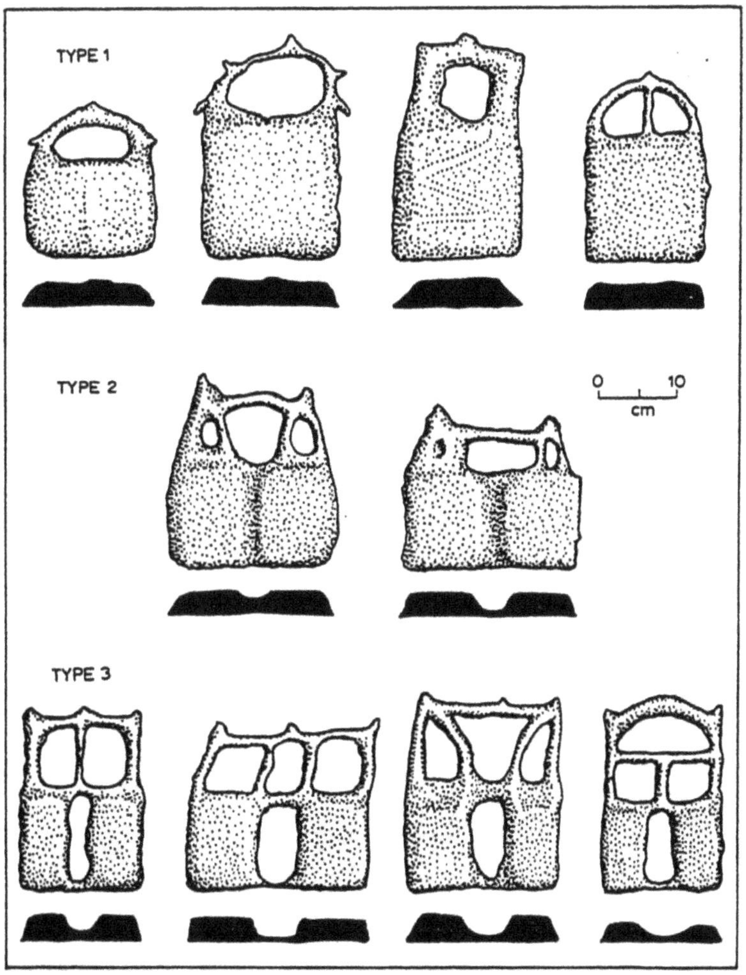

13 Selection of tin ingots showing the variety of forms recovered from the mid-1st century AD wreck in the outer harbour of Port Vendres (redrawn from *Gallia*, 1975, 33, 64)

Fig 3.2 Tin Ingots Mid 1st Century Wreck, Vendes, France, Pannhallurick, R.D., 1986, 107

j) fourteen very pure cast tin ingots, (FIG 3.2) the heaviest one weighing 8.75 kg; all of the ingots carry varying stamps;
[shipwreck by Port Vendes, South Mediterranean coast of France, near Spanish frontier: cira first century AD] (Penhallurick, 1986, 107-9).

k) thirty two tin ingots weighing four-five kg each, stamped with a Latin inscription;
[shipwreck off the east coast of Sardinia Capo Bellavista - Punta Nera: circa first century AD] (Penhallurick, 1986, 109).

It maybe seen from the above list that the examples of pure tin objects and ingots found in an archaeological context are a rare occurrence. The lack of pure tin objects was probably due to the fact that most tin in antiquity was used in the production of bronze. Also the above list shows that the bulk of tin found from antiquity has been found in ingot form.

Tin being considered a rare and very valuable commodity in antiquity, meant that it had a high trade value and would not have been wasted. The rarity of tin presented the problem of supply not meeting the demand; and this necessitated a search for an alternative to bronze, which iron and steel finally became.

Tin was used in bronze body armour by certain sections of the Roman Legions, in everyday household items such as pewter table ware which had a high content of tin (50% Pb and 50% Sn) and in the casting of bronze sculptures (Tylecote, 1992, 72).

Another reason for the rarity of tin in archaeolmetallurgical smelting sites and excavations could be the basic unstable nature of tin. The few pure tin objects and ingots which have been found always have a fairly high level of tin corrosion present.

Corrosion of pure tin can completely destroy the object and any original shape of that object and reduce it to a totally unrecognisable powder. The main corrosion product of tin within a marine environment is tin oxide ($Sn0_2$) which is represented after complete corrosion by a white paste-type stain in the soil.

Pure smelted metallic tin is only stable above 13^0C and is the only common smelted metal which undergoes such a crystalline change as an allotropic transformation from a metal (beta) to a non-metal (alpha) form at normal temperatures (Tylecote, 1986, 49-70). This leads to the possibility that any buried tin metal from antiquity exposed to temperatures below 13^0C for long periods of time would "disappear" and further water leaching of the corroded product tin oxide would almost leave no trace of the original tin object or the corrosion which it had produced (Cronyn, 1990, 212).

The transformation or corrosion of tin can though in some cases be slowed down by small amounts of impurities present in the tin or by the addition of other minerals to form a composite metal during the smelting process (i.e. lead which with tin forms the metal pewter) (Cronyn, 1990, 211).

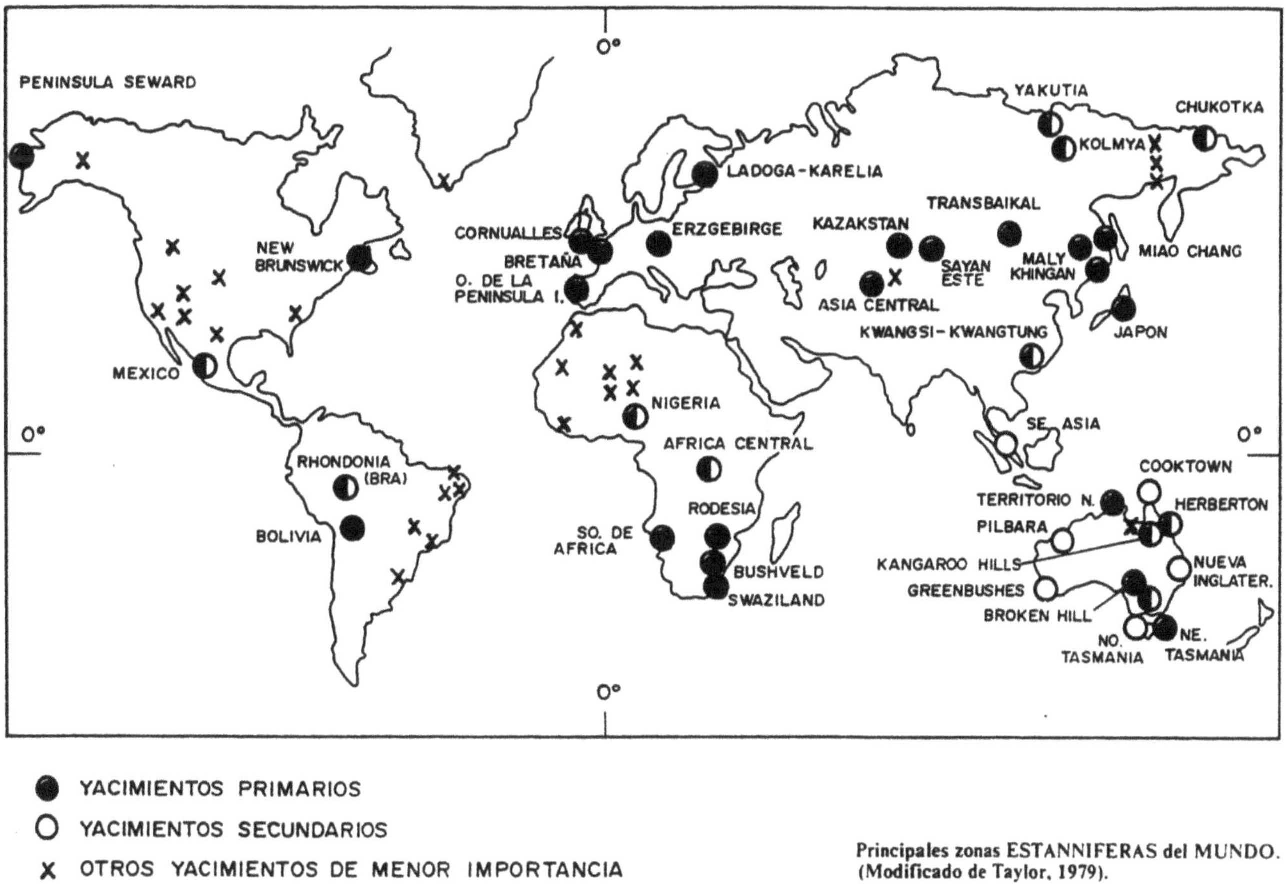

Fig 4.1 TIN DEPOSITS OF THE WORLD, Junta de Castilla y Leon; 1988, 26

This chapter will offer a brief introduction to the formation, geology and types of tin minerals and of tin's relationship with granite.

Tin does not appear in large quantities on the earth's surface. It is present in the Earth's crust at only 3 parts per million. Uranium appears at 2.5 parts per million. Copper appears as 70 parts per million and zinc at 80 parts per million. Silver appears at .08 parts per million; Gold appears at .004 parts per million (Slater, 1974, 3).

The mineral cassiterite SnO_2 has been and still is the main source of tin. In the past cassiterite has been mined from both vein and alluvial deposits. Cornwall and Devon in England are one of the few sources of mined "vein tin" in the world.

The only other country to mine vein
tin is Bolivia in South America. All other tin producing countries have produced and are producing alluvial tin which accounts for approximately 80% of the total world production (Robertson, 1982, 1).

Cassiterite, which is a stable tin oxide, that is both very dense (specific gravity 6.6-7.1) and hard (hardness 6-7 using Mohs' hardness standard) (Atkinson, 1985, 3). The colours of cassiterite are primarily brown or black but in some cases can be a yellowish brown or reddish black. The cassiterite crystal system is tetragonal, normally appearing short and square in section. In some cases, the crystals appear long and thin. In this appearance cassiterite is sometime called "needle tin".

Elluvial tin is weathered from vein cassiterite and is found either lying just below the original vein deposit area or has been moved downhill by weathering or water erosion (Bates and Jackson, 1990, 211). Elluvial tin is tin ore found in any location other than tin veins or stream beds.

Stream tin or alluvial tin was orginally vein cassiterite ore which has weathered out of its granite or quartz host rock over geological time and been deposited in stream and river bed gravels. Most stream cassiterite has been worn into rounded stream pebbles of varying sizes. Stream tin is primarily identified by its heavy weight, high density and black or brown colours.

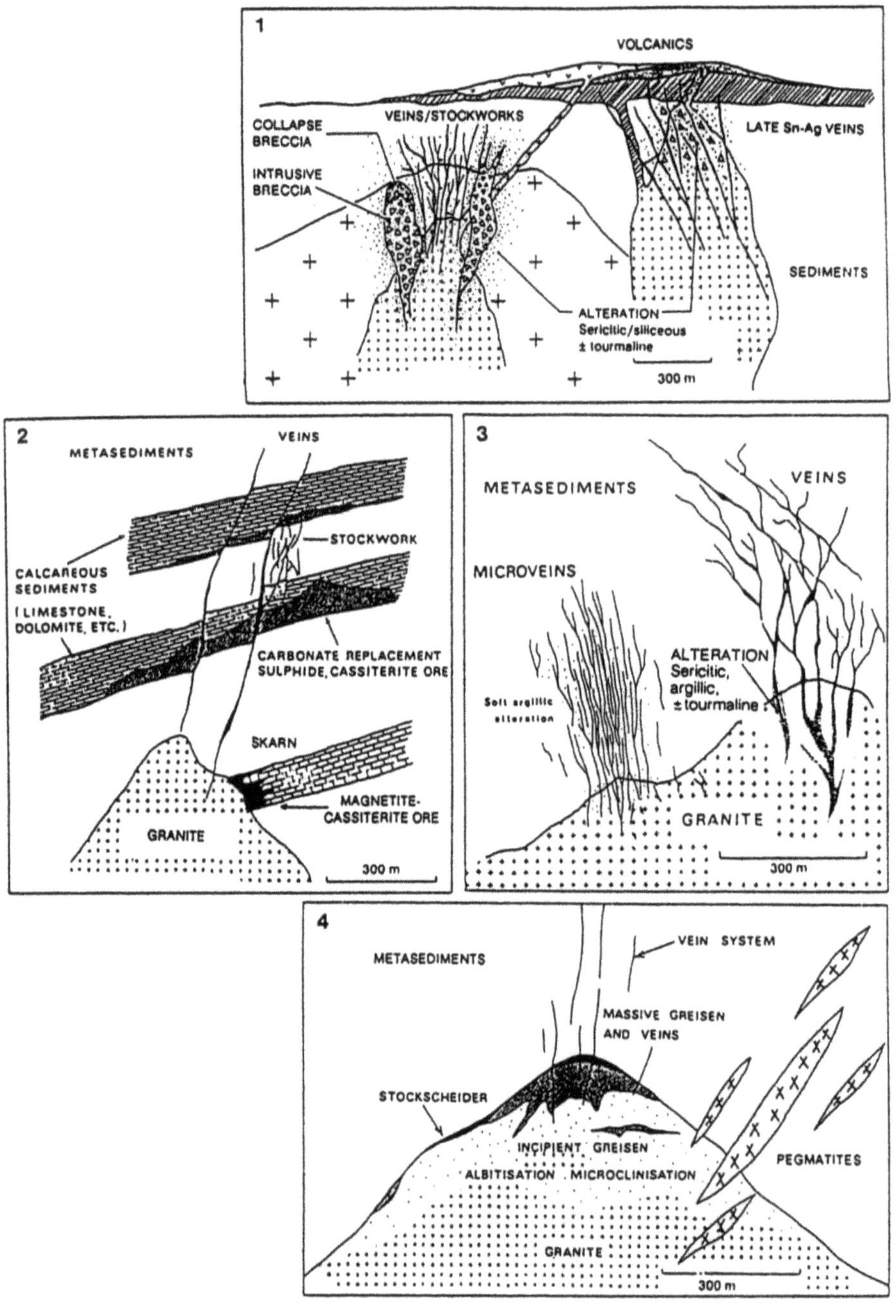

Fig 4.2 Schematic Models of the Major Groups of Primary Tin Ore Deposits after Taylor et al 1985
1) tin porphyries
2) skarns and carbonate/sulphide replacements
3) veins and sheeted veins
4) greisens and pegmatites
Lehmann, B., 1990, 6-7

Cassiterite also sometimes occurs in a form called "wood tin". Wood tin is a brownish coloured form of cassiterite which comes normally in nodular form and has an internal concentric structure of radiating crystals which resemble thin fibres and looks like dry wood (Bates and Jackson, 1990, 714).

Tin also appears as stannite (Cu_2FeSnS_4). Stannite is also sometimes called "bell metal ore". This is a complex sulphide tin-copper-iron ore which is found at deep levels within copper tin vein mines in Cornwall. The levels at which the stannite appears are below the water table and the mines need good pumping equipment before the stannite can be mined economically (Dines, 1988, 20).

Wood tin is also called "toads eye" when it has been formed on a small scale (Cook & Kirk, 1991, 57). It is thought by some people that wood tin is a form of disintegrated stannite Cu_2FeSnS_4 (Penhallurick, 1986, 1).

Wood tin is not as common as the other two forms of cassiterite and would probably not have been used in antiquity (Dines, 1988, 20). It seems that it did appear near the surface in Cornwall. This would have occurred with the weathering out of tin veins which contained some wood tin. Therefore wood tin could have been easily available to ancient man, but it did not appear in large amounts. It is possible that the majority of the easily available wood tin had been collected and used by ancient man.

I cut and mounted a wood tin mineral sample and then analysed it on a Scanning Electron Microscope (Sample No 72). It contained from 50-76% tin (depending on the area analysed) and small traces of iron 0.48% - 1.3%.

Due to the depth and complexity of stannite ore it is very doubtful if it was used as a source of tin or bronze in prehistoric times. It is thought by some geologists because of stannite's placement below the tin deposits in Cornwall that is has been formed through a leaching out process of the higher cassiterite mineral (Penhallurick, 1986, 1).

Tin is almost always found in or near granite and it is also capable of appearing in some contact metamorphic rocks surrounding granite intrusions (Pellant, 1992, 81). Tin does not seem to appear in any other formation context.

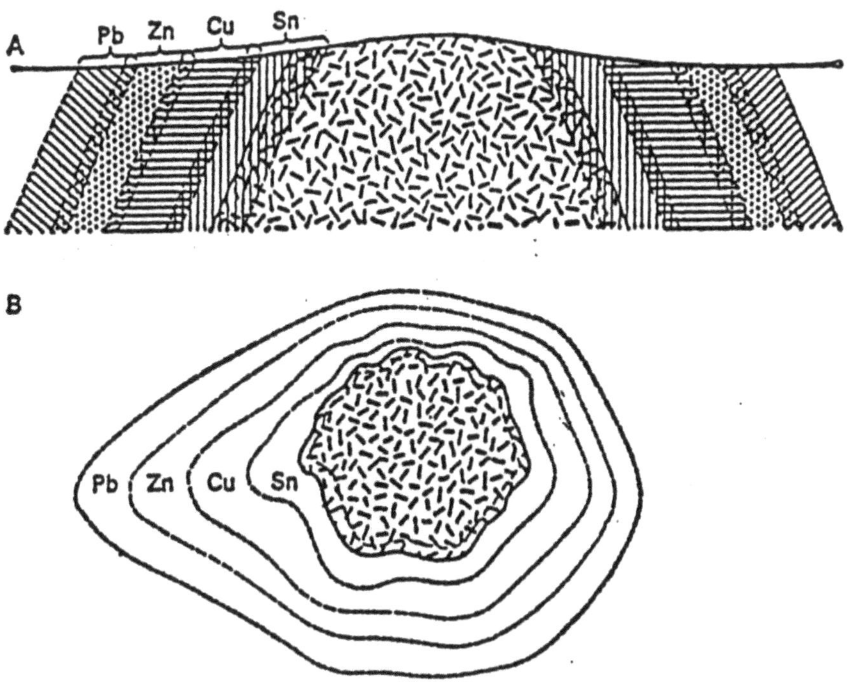

The zoning of mineral deposits around an igneous intrusion.
(A) Section; (B) Plan.

Fig 4.3 Zoning of Mineral Deposits around an Igmeous Intrusion
Whitte, D.G.A. and Brooks, J.R.V., 1972, 231

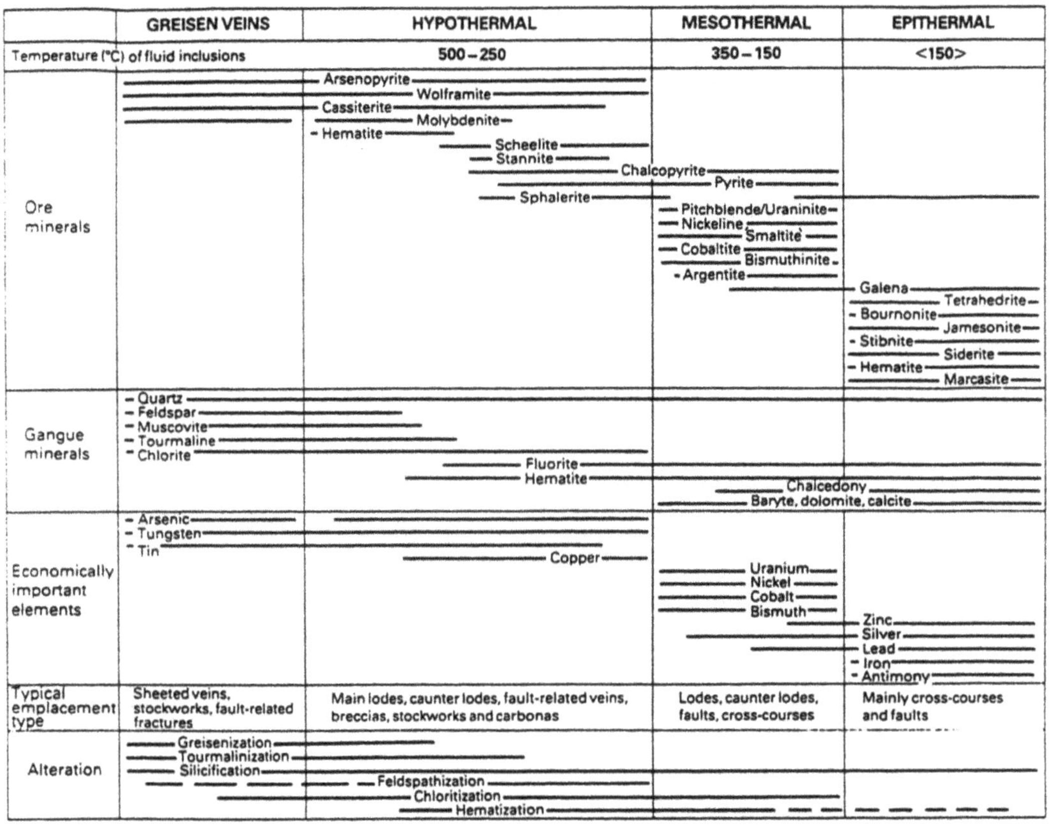

Fig 4.4 Generalised Paragenetic Sequence of Ore and Gangue Minerals
South-West England (after Stone and Exley 1985)
Embry, P.G. and Symes, R.F., 1987, 8

The granite family of rocks are very wide spread rock and found under many different conditions on the Earth's surface. It is difficult to give a general and simple definition of granite that will conform to all the varying granites of the earth's surface.

The name of granite comes from the Latin word ("granum" = grain) and pertains to the granular texture of granite. Granite is a plutonic igneous rock which is coarse grained and consists of quartz feldspar and biotite. The complete granite rock family is the most common of all intrusive igneous rocks (Watt, 1982, 76).

Granite is generally a light coloured rock; this is due to its general composition of which the light coloured minerals quartz and feldspar make up between 80-100%. The light coloured feldspars include the following, alkali feldspar 35-100% and plagioclase feldspar 0-65%. The darker materials which generally make up granite account for only 0-20% and consist generally of the following components biotite, augite, hornblende, muscovite, apatite, zircon and magnetite. These darker minerals which are included in granite never make up more than 20% of the total (Schumann, 1992, 200).

In the earlier years of the study of geology, it was thought that all granite was formed from magma coming to the earths crust from the interior of the earth in a variety of different places and under different physical and chemical influences. This theory is still accepted as basically true, but with the following qualifications.

It is now thought that granite can also have developed from the remelting and reforming of existing crustal rock which has been involved in plate tectonics and metamorphism (Pitcher, 1993, 29-40).

The current theories on the nature and origin of granite are still being debated. It is difficult to reconstruct the formation of granite in a laboratory environment due to the supposed very high pressures and temperatures under which granite is formed. To try to duplicate these pressures and temperatures in the laboratory is at present very difficult (Pitcher, 1993, 39-40).

Granite appears in the following manifestations: dikes, sills, plugs, bosses, ring complexes and when in its largest size as batholiths (Whitten and Brooks, 1972, 213). All cassiterite could have been formed in association with any one of the above granite structures.

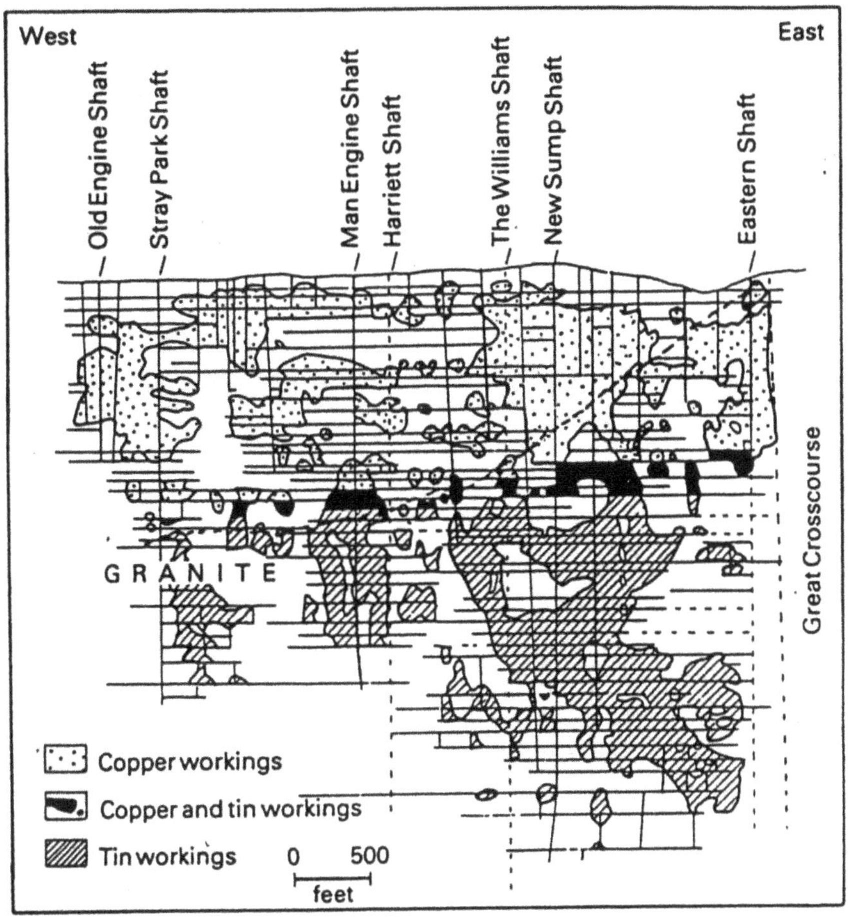

Fig 4.5 Dolcoath Mine Section - Ore Zones / Dolcoath Mine North of Camboren, Cornwall
Embry, P.G. & Symes, R.F., 1987, 13

Cassiterite seems only to appear in the uppermost granite bodies which develop out of lower and larger batholiths. The granites which give rise to the tin minerals are normally rich in quartz, tourmaline and fluorine bearing materials (Lehmann, 1990, 4-6).

One theory on the formation of cassiterite proposes that both cassiterite and wolframite (FeMn)WO_4 are concentrated and formed as a result of pneumatolytic action ("alteration of a rock or crystallisation of minerals by gaseous emanations derived from solidifying magma") (Bates and Jackson, 1990, 513).

"The theory states that the cassiterite and wolfram are moved as fluid fluorides (SnF_4 and WF_6) and react at times with water to form the oxides cassiterite (SnO_2) and wolfram (WO_3). The Wolfram will also either react with Fe and Mn to produce Wolframite (Fe,Mn)WO_4 or with Ca to form scheelite ($CaWO_4$)" (Whitten and Brooks, 1972, 354).

Cassiterite veins are formed in highly fractured acid granite through hydrothermal or pneumatolytic action. It is thought that the cassiterite is formed by fluid evolution during the time that the granite magma is in its final cooling phase after rising up through the earth's crust. The local geological conditions play a very important part in tin's mineral development and placement (Fig 4.2).

Pegmatites are very coarse grained intrusive igneous rocks. They normally appear as deformed dikes, lenses or veins. Pegmatite veins rise from the upper surfaces of lower and larger granite bodies such as batholiths and are one of the main carriers of tin minerals. Pegmatites do not normally reach the earth's surface, but are exposed by the process of erosion (Fig 4.2).

Pegmatites represent the final high temperature fluid crystallisation of a magma body. They therefore normally contain much higher amounts of metaliferous minerals than the lower granite body itself, which cooled at lower temperatures (Bates and Jackson, 1990, 488).

Cassiterite appears sometimes with Wolframite, Scheelite, and Arsenopyrite. These minerals are not normally in the same locality as tin, but can either be in a different lateral or horizontal placement; either higher or lower or further away from the cassiterite depending on the temperature at which they crystallized from the magma mineral fluid.

Wolframite the mineral which most commonly appears with tin is normally found at higher levels in the host rock than cassiterite. In Cornwall, copper minerals appear also with cassiterite but much closer to the surface,(FIG 4.5) while the cassiterite remains in the lower part of the ore body (Dines, 1988, 8).

Cassiterite develops and crystallises from its fluid magma host between 600 and 500 centigrade (Sos Baynat, 1967, 426). It is the different temperatures at which the other minerals accompanying tin also crystallises that define their geological and placement relationship to each other (FIG 4.4).

Cassiterite is normally found in quartz veins located within the granite host The cassiterite veins within the quartz vary in width and length depending upon the conditions in which the vein was formed. The sizes of the cassiterite containing veins can vary considerable from 25mm to over a meter in width. The cassiterite veins can in some cases reach lengths of up to 800 meters (Atkinson, 1985, 3).

The gangue material normally accompanying cassiterite is quartz and tourmaline (Dines, 1988, 20).

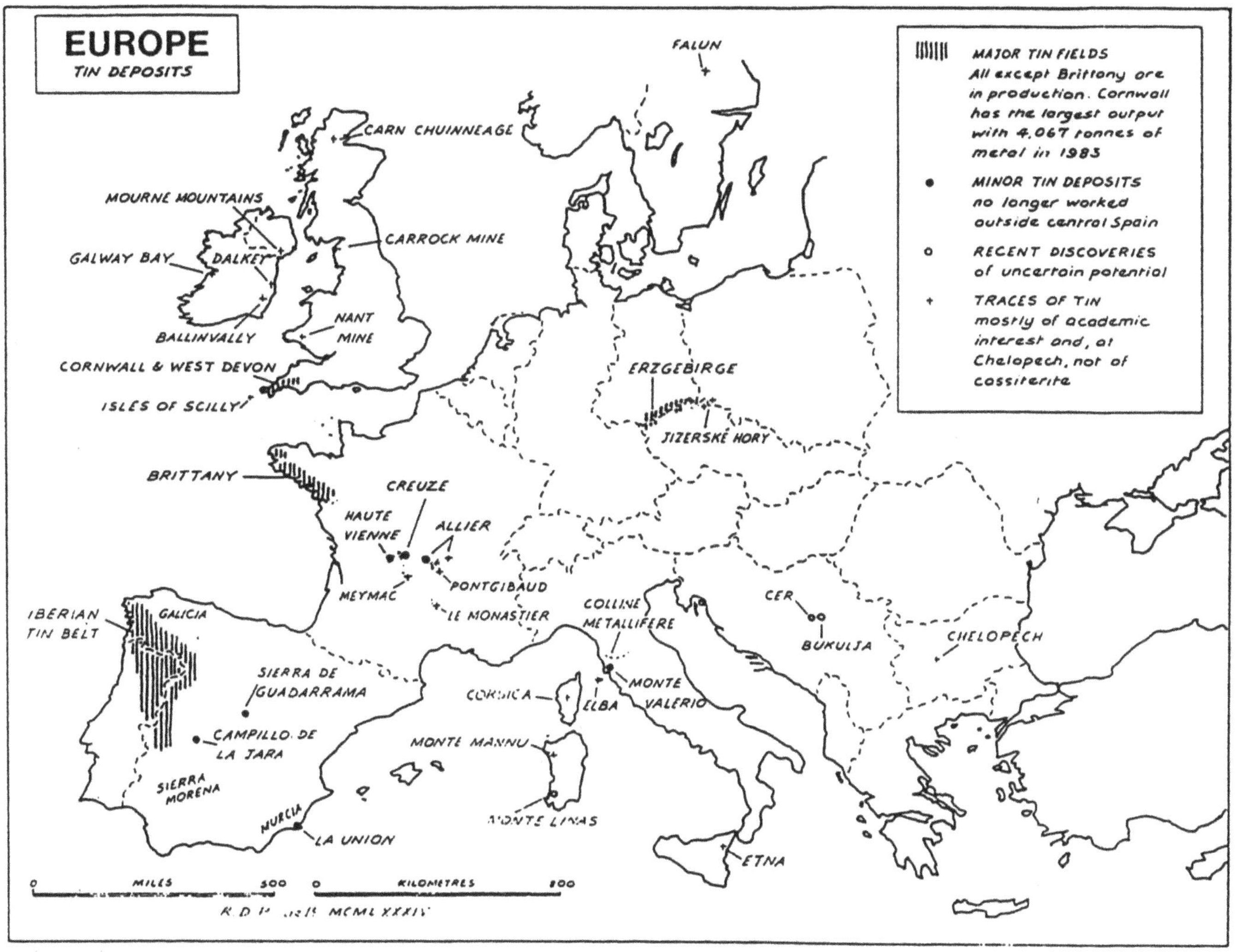

Fig 5.1 EUROPE TIN DEPOSITS
Penhallurick, R.D., 1986, 63

Tin does not appear geologically within many areas of Europe, but where it does appear in commercially viable deposits it is only found within Hercynian granite. The Hercynian Orogeny was a geological time of mountain building which took place in Europe during the late Paleozoic Era lasting through both the Carboniferous and Permian periods (Fig 6.1) approximately 360-245 million years ago (Bates and Jackson, 1990, 304). The Hercynian mountain building period is also referred to in some Geological publications as the Variscan Orogeny (Fig 5.4).

It should be noted that any granites formed in the later Alpine Orogeny during the more recent Tertiary era 65-3 million years ago, have not produced to date any geological evidence of large commercially viable cassiterite deposits.

There are four main tin bearing areas in Europe which have been worked in the past or are still being worked at present but on very reduced levels (Fig 5.1). There are also two smaller tin mineralization areas in Europe: The Tuscany area of Italy and the island of Sardinia where tin might possibly have been mined in antiquity (Penhallurick, 1986, 79-80).

The most northerly of the four main European tin mineral areas is the granite area of Cornwall and Devon in the Southwest peninsula of England (Embury and Symes, 1987, 2-14). This rather compact rich tin area is approximately 30 kilometres wide by 150 kilometres long. It is believed and much quoted that the tin deposits of Cornwall and Devon were worked in prehistory, but to date archaeological evidence supporting this theory is almost non-existent (Ager, 1980, 167).

In France there are two tin areas. The granite area of the Armorica Massif which covers all of the Brittany peninsula (Fig 5.2). The tin appears mostly in the Western (Department of Finistere) and Southern areas (Departments of Morbihan and Loire Atlantique) of Brittany; with only a small scattering of tin mineralization areas in the northern and eastern areas of Brittany (Penhallurick, 1986, 85-94).

The second area within France is the tin deposits located on the Northern edge of the Central Massif in the departments of Haute Vienne, Creuse and Allier.

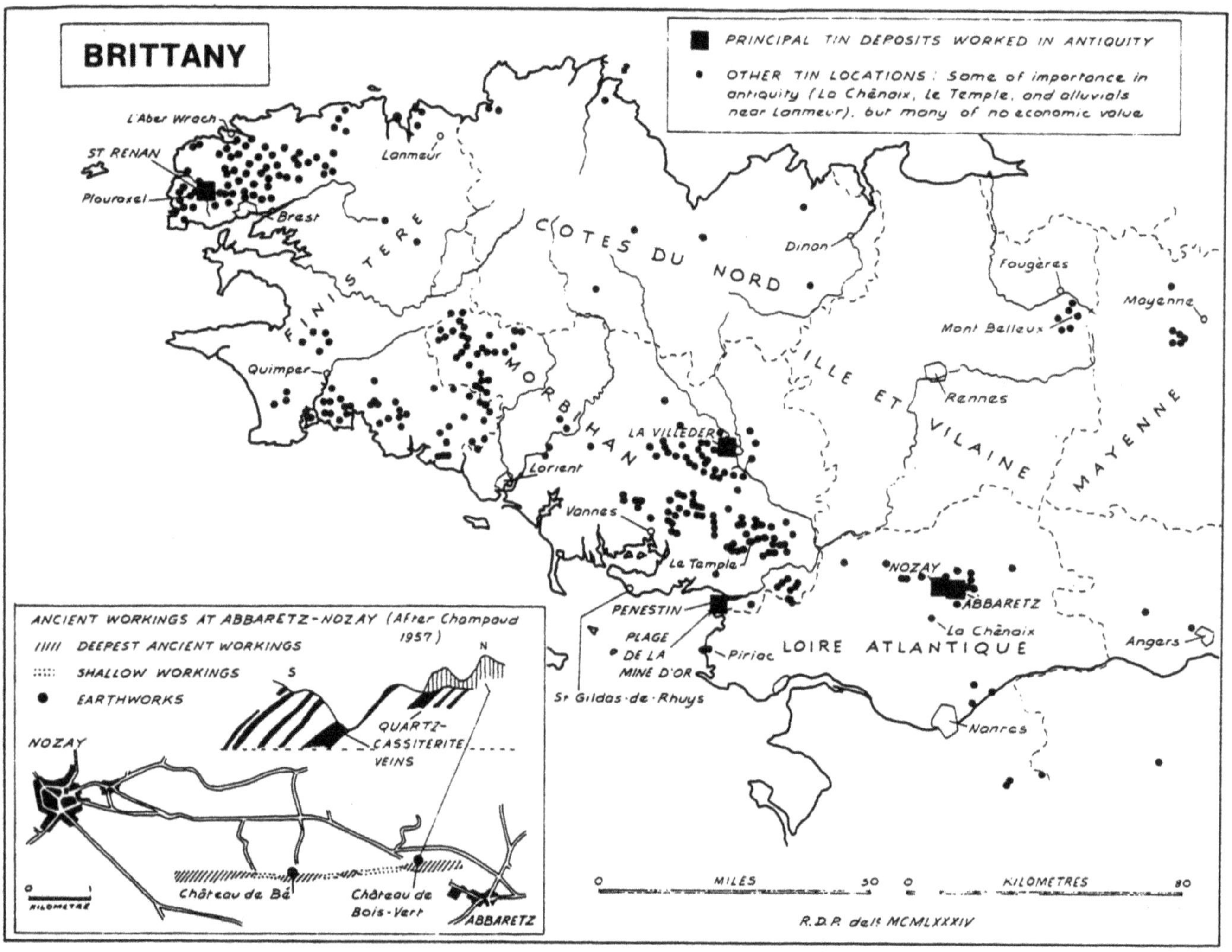

Fig 5.2 Brittany - Principal Tin Deposits, Penhallurick, R.D., 1986, 87

The tin areas of the Central Massif have not to date offered any archaeological evidence tying the tin deposits to use by ancient man (Penhallurick, 1986, 85-6).

In the south/east area of Germany and the north/west area of Czechoslovakia there is a third major cassiterite area (Fig 5.3). This area is called the Erzgebirge (Ore Mountains). The Erzgebirge area is located within the Bohemian Massif in the border region between the German land of Saxony in the Southeast corner of Germany and the North Western area of Bohemia in the Czechoslovakia Republic.

The Erzgebirge have a very long and rich history as a mining region and right into the Middle Ages the region was still a major tin producer. It is this mining area on which Georgius Agricola wrote his classical work De Re Metallica in 1556 (Penhallruck,1986, 71-83). To date firm, archaeological evidence linking tin production in the Erzgebirge with antiquity and early metallurgical periods,is still lacking.

It was approximately 125 kilometres to the Southeast of the Erzgebirge in the region of modern day Prague that the Bronze Age Unetice culture (1900 BC - 1500 BC) was centred with its large scale bronze metal working industry. The Erzgebirge tin area represents the only plausible mineral location for the tin used to produce the Unetice bronze archaeological artefacts. It is possible that any evidence of ancient tin mining within the Erzgebirge region has been destroyed by later mining activity both Medieval and modern, or has just been generally overlooked and ignored; possibly a bit of both.

The fourth and final major area within Europe where tin appears in notable amounts is within the granites of the Iberian Massif which geographically covers parts of both Spain and Portugal. The Iberian Massif covers about two thirds of the Iberian Peninsula, running from the province of Galicia in the Northwest corner of the Iberian Peninsula to the Guadalquivir River valley fault in Andalusia.

The granites of the Iberian Massif contain the majority of the tin deposits both 'mined out' and in current production within the Iberian Peninsula.

There are or were a few small mineral deposits of tin outside the Massif; San Isodoro by Cȁrtagena and Marinera near Almeria in the Betic Cordillera (Davies, 1919, 84). The Betic Cordillera forms the most Western end of the Alpine Orogenic belt, which was formed during the Tertiary Era (65-3 million years ago).

The above mentioned areas which are located in the provinces of Murcia and Amleria have no real commercial value for modern mining enterprises. It is possible that both of these areas might have been worked in antiquity. At present there is no archaeological proof to support this hypothesis. The above two areas could sometime in the future be surveyed for ancient mining and metallurgical evidence.

It is assumed due to geological similarities that the areas of Cornwall, Devon, Brittany and the Iberian Massif were in very close proximity during some of their geological formation time. They have moved to their present positions due to the forces of plate tectonics (Ager, 1980, 222).

The main mineralization is quite similar in all three granite areas. The single difference being that Niobium, Tantalum and Titanium show up regularly with tin (or in very close proximity) in the Iberian Massif, but do not seem so prominent in Cornwall or Brittany (Dines, 1988, 20-32)

The final two mineral tin areas which need to be assessed are the Mediterranean areas of Tuscany in Italy and the island of Sardinia

The very fertile and metal rich region of Tuscany, Italy, is centred in the Anti-Apennines mountains on the Western side of the Italian Peninsula; bordered by the Arno river to the North and the Tiber river to the East and South (Grant, 1980, 9-16). The Tuscany region was the core area of the Etruscan city state confederation 8th to 5th century BC.; which had a major influence on the development of the Roman culture (Whitehouse, 1983, 166-7).

At present the only known tin deposit of the Tuscany region is at Monte Valerio close to the village of Campiglia Marottima, about 10 kilometres from the coast. The mineralization of the Monte Valerio tin deposit consists of cassiterite scattered throughout a larger brown hematite iron bearing deposit. Also present are small amounts of sphalerite, galena, and chalcopyrite.

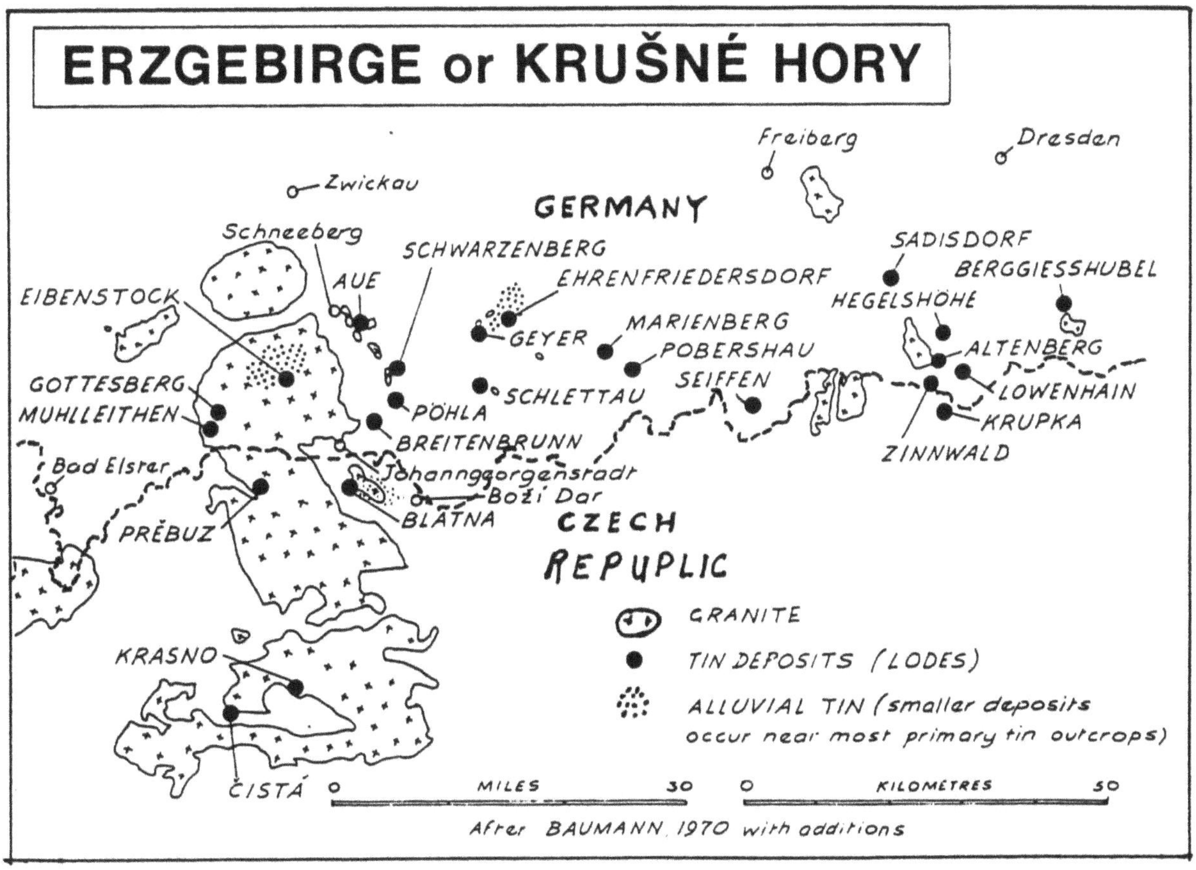

Fig 5.3 Erzgebbirge or Krusne Hory, tin area, Penhallurick, R.D., 1986, 72

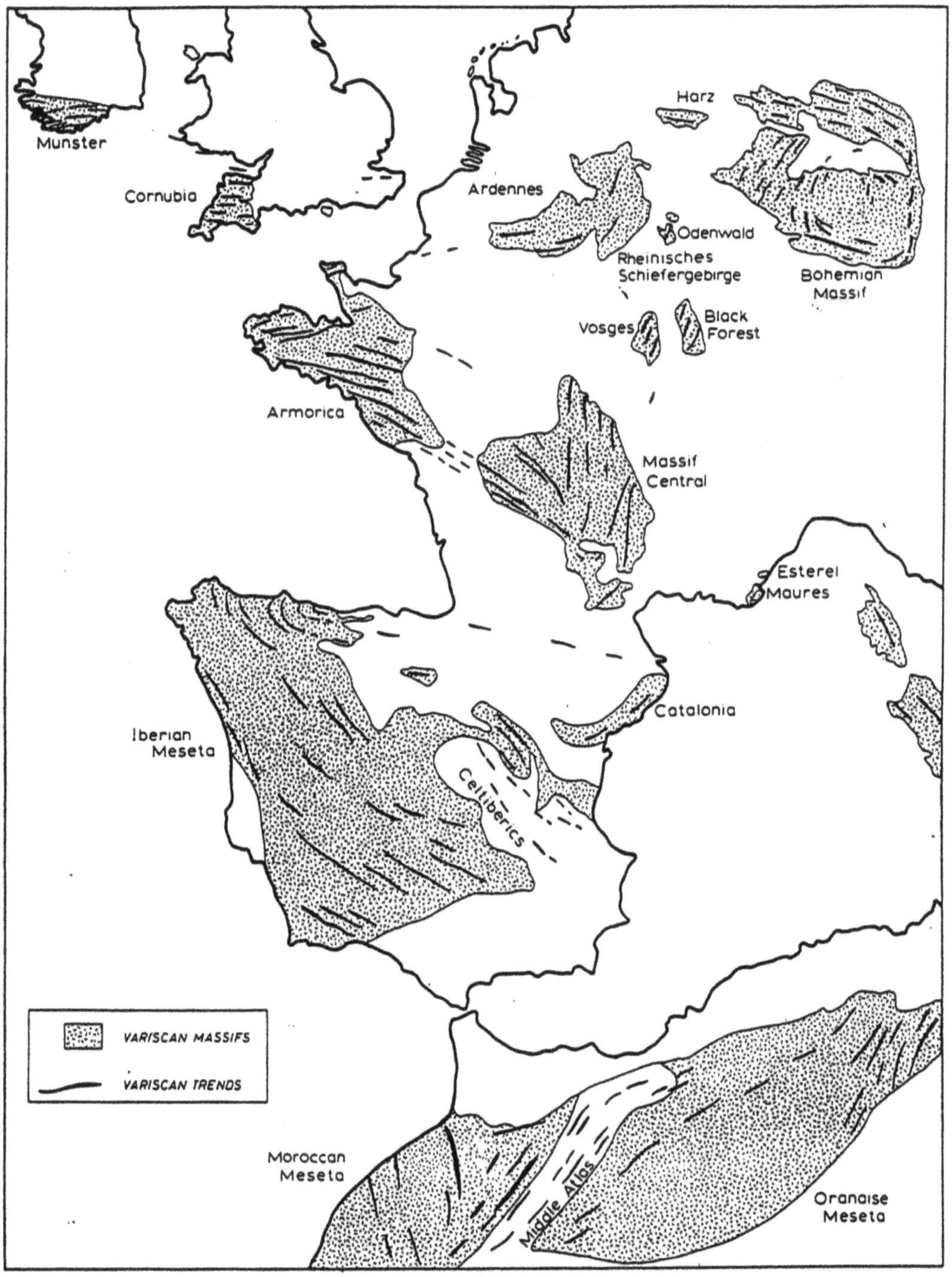

General trends of the Variscan massifs of western Europe (from Ager, 1975, by kind permission of the Council of the Geologists' Association).

Fig 5.4 General Trends of the Variscan Massifs of Western Europe
Ager, D.V., 1980, 314

The geological formation of the Monte Valerio tin deposit consists of a low temperature granite pluton intruded in limestone in the Pliocene Epoch approximately 45 million years ago; making it one of the most recent tin deposits in Europe (Penhallurick, 1986, 80).

Possible ancient workings have been found at Monte Valerio, but it can not be proven what period of antiquity they represent and if they were centred around the tin ore deposits or the much larger iron ore deposits of the region. In any case the tin deposits of Monte Valerio were not considered to have been very extensive and could have only provided small amounts of tin in antiquity (Penhallurick, 1986, 80).

The other tin area is in "Sardinia which has had a very rich archaeological history from the beginning of the third to second millennium BC" (Guido, 1963, 28). Deposits of cassiterite are to be found in small amounts on Sardinia. Sardina however is a richly mineralized Mediterranean island which has major mineral deposits of Cu, Pb, Ag, Fe, Zn and Ni.

One tin mineral area is located in the North/West of the island near the town of Alghero. This deposit consists of a complex ore containing cassiterite, sphalerite, chalcopyrite, iron pyrite, and galena. A second tin mineral area is located in the South/West of the island in the district of Iglesias. This second deposit occurs in quartz veins (Penhallurick, 1986, 79-8). It is believed due to lack of any earlier evidence that the above two cassiterite deposits had not been worked before the Phoenician and Roman periods 500 BC (Guido, 1963, 28).

At present it is believed that any tin used in bronze production on Sardinia in antiquity was brought in by ship from either Tuscany in Italy or from the iberian tin mineral regions (Penhallurick, 1986, 8).

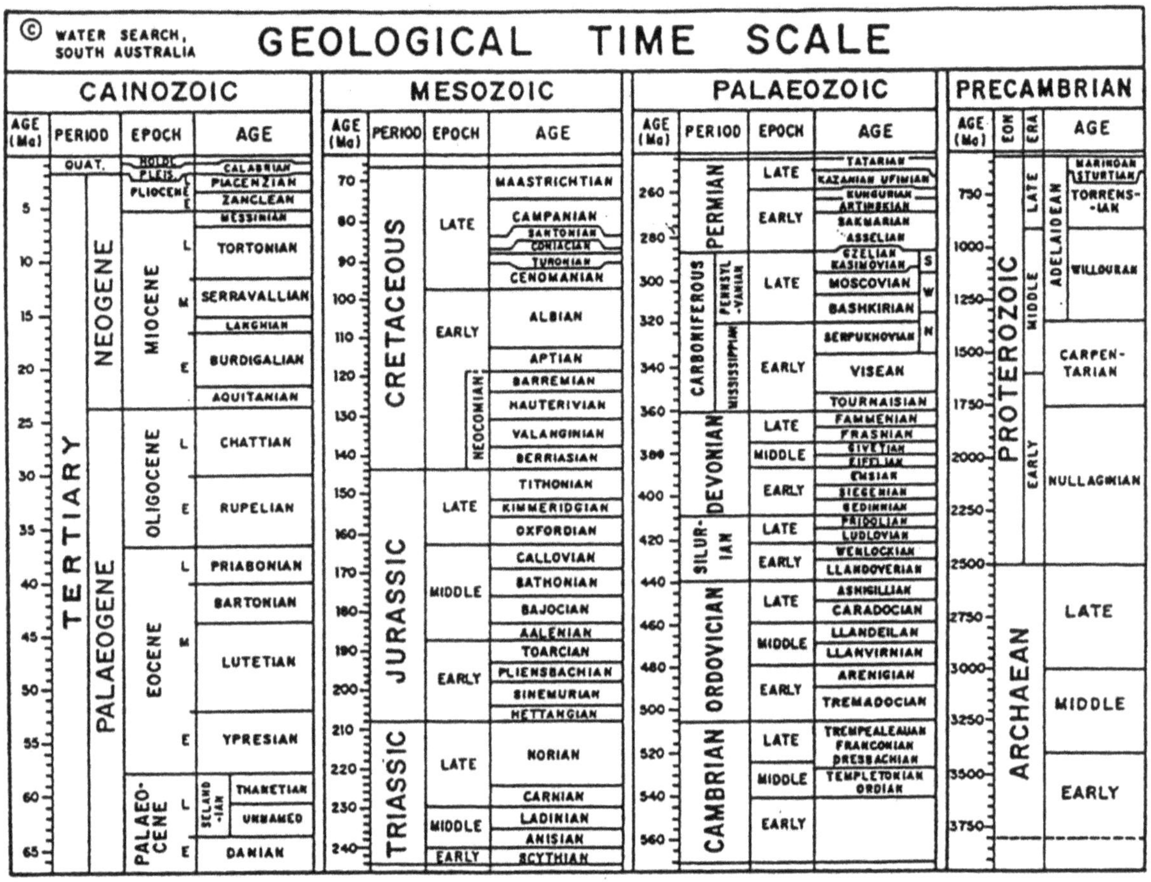

Fig 6.1 GEOLOGICAL TIME SCALE
Water Search - South Australia

The main area in which I conducted my survey work in both Spain and Portugal was within the Iberian Massif. The geological interpretation of the Massif is based around two slightly differing theories. The first or more recent theory (developed between 1926 to 1988) states that the majority of the rocks which make up the Massif are of Archean (older Pre-Cambrian) age.

The Archean rocks were then reworked during the Caledonian Orogeny, a mountain building period, that took place during the Silurian and Devonian periods 435 - 360 million years ago. The Massif was then reworked again during the Hercynian Orogeny, which took place during the Carboniferous and Permian periods 345 - 230 million years ago (Dallmeyer and Garcia, 1990, 3).

The second and slightly older theory (developed between 1924 to 1945) states that the Massif was only formed during the Hercynian Orogeny and had no connection with any earlier rock formations or mountain building (Dallmeyer and Garcia, 1990, 3).

What both theories then agree on is that the Iberian Massif represents the most western mountain development of the European Hercynides mountain systems and that this very large area was not geologically modified by the Alpine Orogeny which reformed large areas of Europe in the Tertiary Era 65 - 24 million years ago (Dallmeyer and Garcia, 1990, 3).

Within the Iberian Massif the later formed rocks are all located at the outer edges and the older rocks of Lower Palaeozoic and Precambrian ages in the central area.

The older Precambrian rocks of the Iberian Massif consist of amphibolites, peridotites, gneisses, eclogites and granites. The younger Precambrian rocks are found in large areas of the Massif and are made up of gneisses, schists, slates, sandstones, quartzites and acid and basic volcanic rocks.

The Cambrian rocks are mostly conglomerate's, slate's, sandstone's, limestone's and volcanic rocks. The rocks of the Ordovician Period consists of slate's, sandstone's, quartzite's and limestone's. The Silurian Period rocks are slate's and sandstone's (Dunning et al., 1990, 108-9).

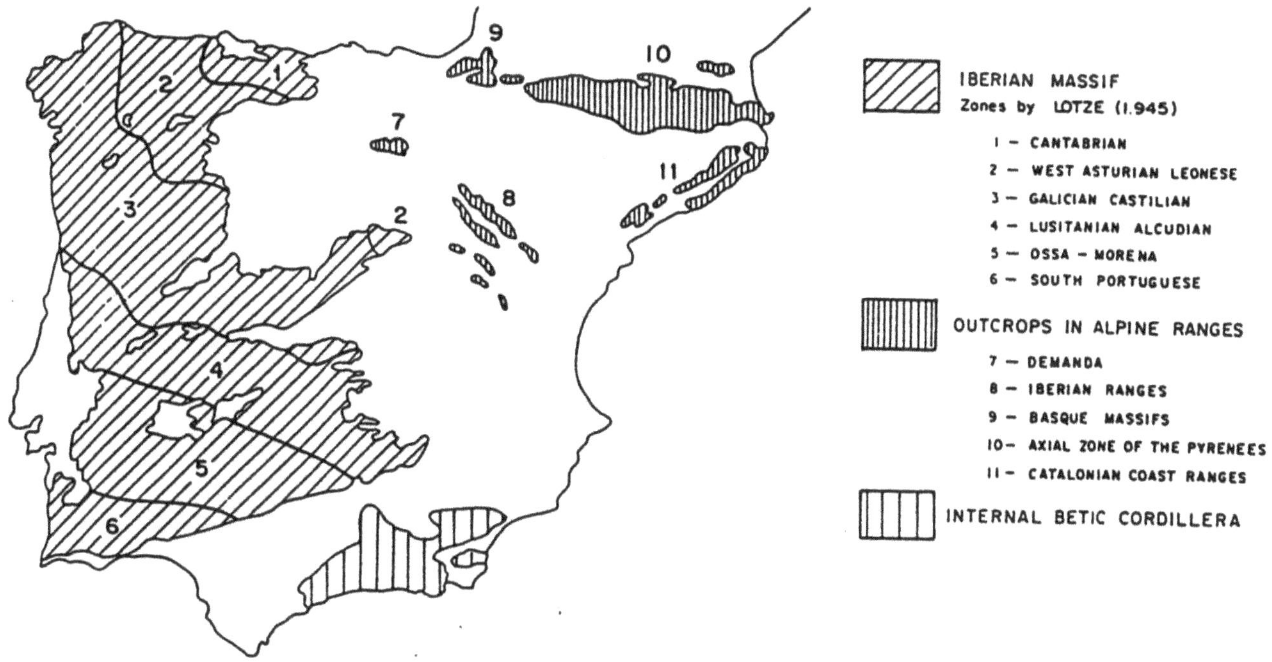

Fig 6.2 Exposures of pre-permain rocks within the Iberian Peninsula
Dallmeyer, R.D. and Garcia, M.E. (ed), 1990, 3

The Devonian rocks which are mostly located in the outer areas of the Massif are slates, sandstones, quartzites and limestones. The Carboniferous rocks are also located in the outer areas of the Massif and comprises an assortment of facies and lithogenesis.

The two Orogenic periods that affected the Iberian Massif were the earlier Caledonian Orogeny which took place during the Silurian and Devonian Periods and only gently folded certain areas within the Massif (Dunning et al., 1990, 199). The later Hercynian Orogeny was a period of intense folding and mountain building which produced wide ranging areas of metamorphism and granitization within the Iberian Massif (Dunning et al., 1990, 109).

In 1945 the geologist Lotze divided the Iberian Massif into six distinct regions (Fig 6.2). He based his divisions on the many stratigraphic and structural variations which appear within the Iberian Massif. His geological regional organisation was slightly altered in 1972 by Julivert et al., (1972) who combined his two central regions (Galician Castilian and Lusitanian Alcudian) into one central region and this arrangement is still in general use today (Dallmeyer and Garcia, 1990, 3-4).

The five regions starting from the Northwest of the peninsula and going down to the Southwest are 1) The Cantabrian and Palentian Zones, 2) The West Asturian-Leonese Zone, 3) The Central-Iberian Zone, (consisting of areas three and four on the accompanying map) 4) The Ossa - Morena Zone (area five on the map) and 5) The South Portuguese Zone (area six on the map) (Dallmeyer and Garcia, 1990, 3-4).

As in most devised Geological systems the argument is still on going as to exactly where one zone starts and the neighbouring zone ends. I am not sure that a final decision on the sub borders within the Iberian Massif will ever be agreed upon by all Iberian Geologists (Dallmeyer and Garcia, 1990, 249).

Zone three, (the combined three and four Zones) called The Central Iberian Zone which is the largest of the five zones, is the area of the Massif in which I did most of my survey work. The Central-Iberian Zone is the middle and most unmodified geological zone of the Hercynian Massif.

208 F. J. Martínez, L. G. Corretgé and O. Suárez

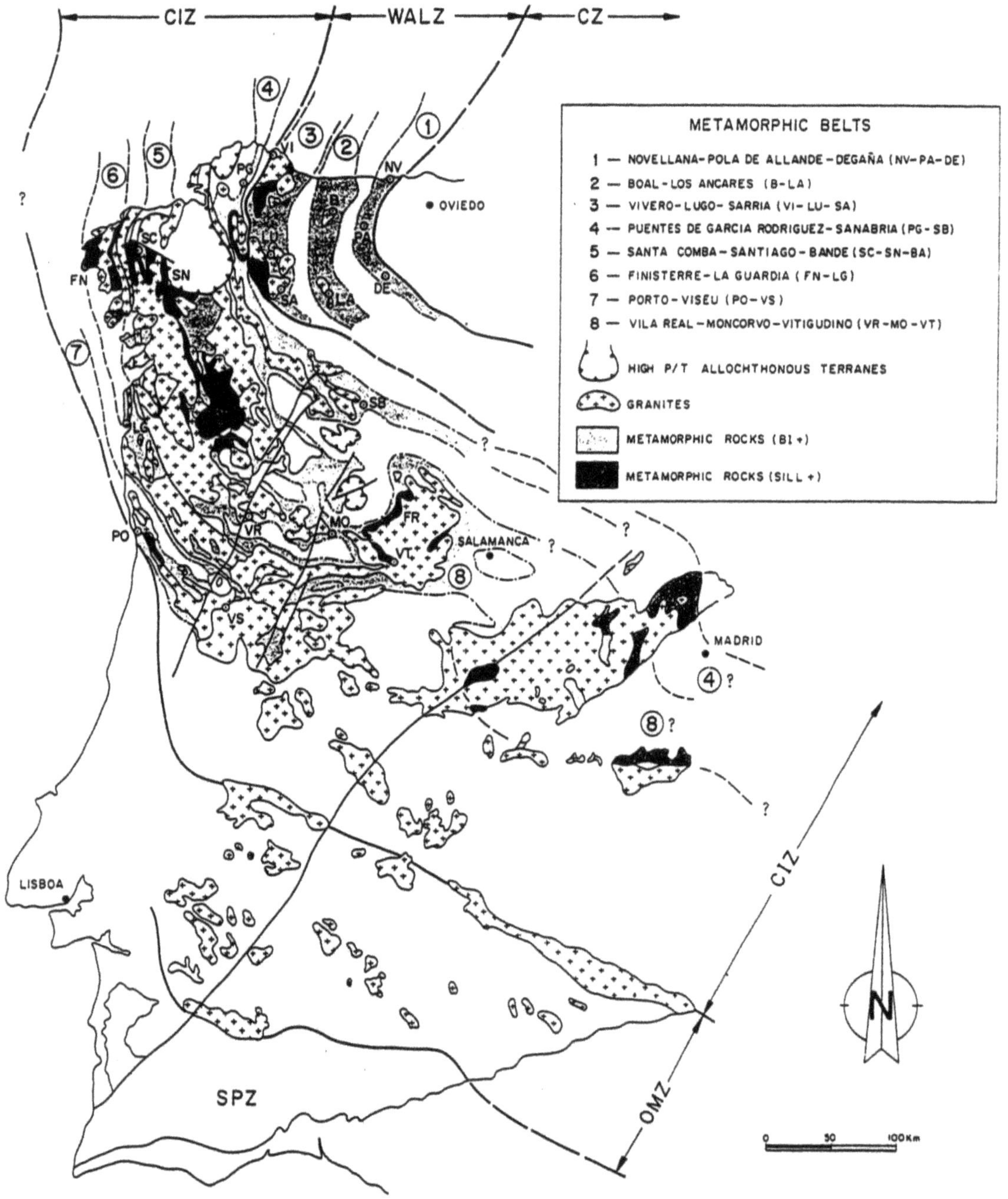

Metamorphic belts distribution in the West Asturian-Leonese Zone and Central-Iberian Zone: *CZ* Cantabrian Zone; *WALZ* West Asturian-Leonese Zone; *CIZ* Central-Iberian Zone (including also Galicia Tras Os Montes Zone); *OMZ* Ossa-Morena Zone; *SPZ* South Portuguese Zone

Fig 6.3 GRANITE AND METAMORPHIC DISTRIBUTION IBERIA
Dallmeyer, R.D. and Garcia, M.E., (ed), 1990, 208

The Central Iberian Zone is one of the most important geological metallogenic areas in the Iberian Peninsula. Included in the Central Zone at the South-eastern end are the famous Almaden Mercury mines and the Pb-district of Linares-La Carolina in the northern part of the province of Jaen. Other mineralizations that appear within this zone in large amounts are Cassiterite SnO_2, Wolframite $(FeMn)WO_4$, Pitchblende UO_2, Autunite $Ca[UO_2/PO_4]_2$ 8-12H_2O, Spodumene $LiAl[Si_2O_6]$, Galina PbS, Sphalerite ZnS, Stibnite Sb_2S_3 and Gold Au (Dallmeyer and Garcia, 1990, 212).

The magmatism of the Hercynian Orogeny (Fig 6.3) led to the development of almost all of the tin and tungsten deposits of the Iberian Massif (Dunning et al., 1989, 109). The following three paragraphs will detail the formation of the three major tin mineralization deposit formations within the Central Zone.

In the first deposit formation the tin generally appears with tungsten. The following description can cover the locations and formation of both the tin and tungsten minerals. The tin mineralization is generally found in veins and stockwork from 1 cm to 1 meter in width and trending mostly in a NE direction; in the vicinity of or actually in exposed granite. Only the following four mined deposits have a much higher percentage of tin than tungsten; Telba and Logorsan in the province of Caceres and Calabor and Arcillera in the province of Zamora (Dallmeyer & Garcia, 1990, 216).

The granite host rocks in the vicinity of the tin-tungsten veins are affected by greisenization (hydrothermal alteration) and tourmalinization (Dallmeyer and Garcia, 1990, 216).

The mineral groups in almost all of the tin/wolframite ore deposits are comprised of the high temperature minerals cassiterite, wolframite and scheelite. There is a transitional lower temperature sulphide stage which is formed in two phases: the first phase is made up of the minerals molybdenite, arsenopyrite, stannite, bismuth, and bismuthinite.

The second phase and final low temperature stage consists of pyrrotite, chalcopyrite, sphalerite, and pyrite. Quartz was produced at all temperatures (Dallmeyer and Garcia, 190, 216).

In the second deposit formation Cassiterite appears with Spodumene in pegmatite deposits. The Sn-Li pegmatites appear in the forms of dikes, and are always found in the vicinity of or in relationship to other varieties of cassiterite mineral deposits.

The following mines are Cassiterite Spodumene pegmatite mines in Pontevedra province the Lalin mine, in Salamanca province the mines of Castillejo and La Fregeneda, in Caceres province the mine of Las Navas and in Badajoz province the mine of Tres Arroyos (Dallmeyer and Garcia, 1990, 217).

The third deposit formation of tin appears in apogranites which have originated during the final process of the magmatic development of mica-rich leucogranite. The apogranites appear as dykes, stocks and sheets and associated granite cupolas. "Apogranites are albitized and greisenized granite located at the perpheral parts of certain intrusions, and are commonly mineralized with one or more of the following rare elements Nb, Ta, Li, Rb, be, Sn, W, Mo" (Bates and Jackson, 1990, 31). The tin comes in very finely dispersed form within the very silica rich areas of the apogranites.

The following mines are apogranite deposits: The Penouta mine in Orense province, the Golpejas mine in Salamanca province and the El Trasquilon mine in Caceres province (Dallmeyer and Garcia, 1990, 217).

Fig 7.1 RELIEF MAP OF SPAIN
LaLaguna J., 1990, 2

The geographical areas where my archaeometallurgical survey took place lie primarily between the upper areas of the Duero, Tagus and Guadiana Rivers and includes regions of both Spain and Portugal, geographically covering part of the Southern Meseta, The Central Sierra Mountains, and the Northern Meseta.

From early antiquity the primary routes for cultural and trade exchange for the Central Mid-Iberian region alternated between either the West and the South. The main natural geographical route to the inland areas of Mid-Iberia from the West was via the Atlantic Coastlands using either the Duero or Tagus River valleys, as the means of approach.

Trade and influence also came from the South by way of the Guadiana River valley; or via overland travel/trade routes from the areas of the modern port cites of Huelva or Cadiz. From Huelva, the ancient roads would have followed approximately what is today the N 435 route to Zafra and there joined up with the present day N 630 route.

From the Cadiz area, the modern N 630 road generally reproduces the ancient route to Mid-Iberia and the North East of the peninsula. The modern N 630 route follows what is considered the major trade route of antiquity running from north to south in the Western side of the Iberian Peninsula, from the Neolithic onwards if not earlier (Galan, 1991, 10).

There must have been some cultural influence and trade coming into the Mid-Central Iberian region via the North and Eastern areas of the Iberian Peninsula, but to date the majority of existing archaeological and cultural evidence points to influence and trade primarily coming and going to from the Western or Southern coastal areas of the Peninsula.

There is evidence of man being present in the Palaeolithic period within the Western Mid Central Iberia area. The evidence is represented by worked stone pebble tools, with some Palaeolithic rock paintings and engravings also present (Ortiz, 1990, 32-5; Gallo, 1985, 109).

The Mesolithic period (9500 BC - 6000 BC) at present does not seem to be well represented in the Mid-Central Iberian Area. The nearest region which has produced abundant evidence of human activity in the Mesolithic is the lower Tagus Valley and the adjoining Atlantic Coastlands areas. The general decline of human activity judged solely by lack of archaeological remains during the Mesolithic period in certain areas of the Iberian Peninsula is put down to climate change.

The major climate change of warmer continental temperature Pre-Boreal and Boreal (8000 BC - 5000 BC) was brought about by the end of the last Ice Age around 8500 BC (Clark, 1952, 12). It is thought that the climate changes, caused by the final melting of the large ice sheets in Europe was responsible for the movement of the majority of the existing game in Iberia into Northern Europe and that the late Palaeolithic hunters moved with the game in search of newer/richer hunting regions (Savory, 1968, 44-6).

The Neolithic (6000 BC - 3500 BC) in the Mid-Iberia area is another period in which the archaeological evidence is not well represented. The closest area with numerous Neolithic archaeological evidence is again the lower Tagus River Estuary and the adjoining North and South Atlantic Coastland regions.

Two existing late Neolithic sites within the region of Extremadura are the Cueva de la Charneca, Termino Municipal de Oliva de Merida and the Cueva de Maltravieso, which had Neolithic artefacts and rock paintings.

Three new major archaeological factors appear in the West Central Mid-Iberian region during the III and II millennia BC, which corresponds generally to the Chalcolithic and the early Copper periods of this region; (1) introduction and use of Megalithic tomb structures, (2) introduction of copper metallurgy used in the form of cast and worked tools and weapons, and (3) appearance of the Beaker material culture complex with its very distinctive pottery and metallurgical objects.

In some regions of Iberia including the area I surveyed, it is difficult at present to understand the order of appearance of the above three archaeological elements. The present archaeological evidence points in some places to all the factors taking place within a very short period of time or even at the same time, but in adjoining geographical regions.

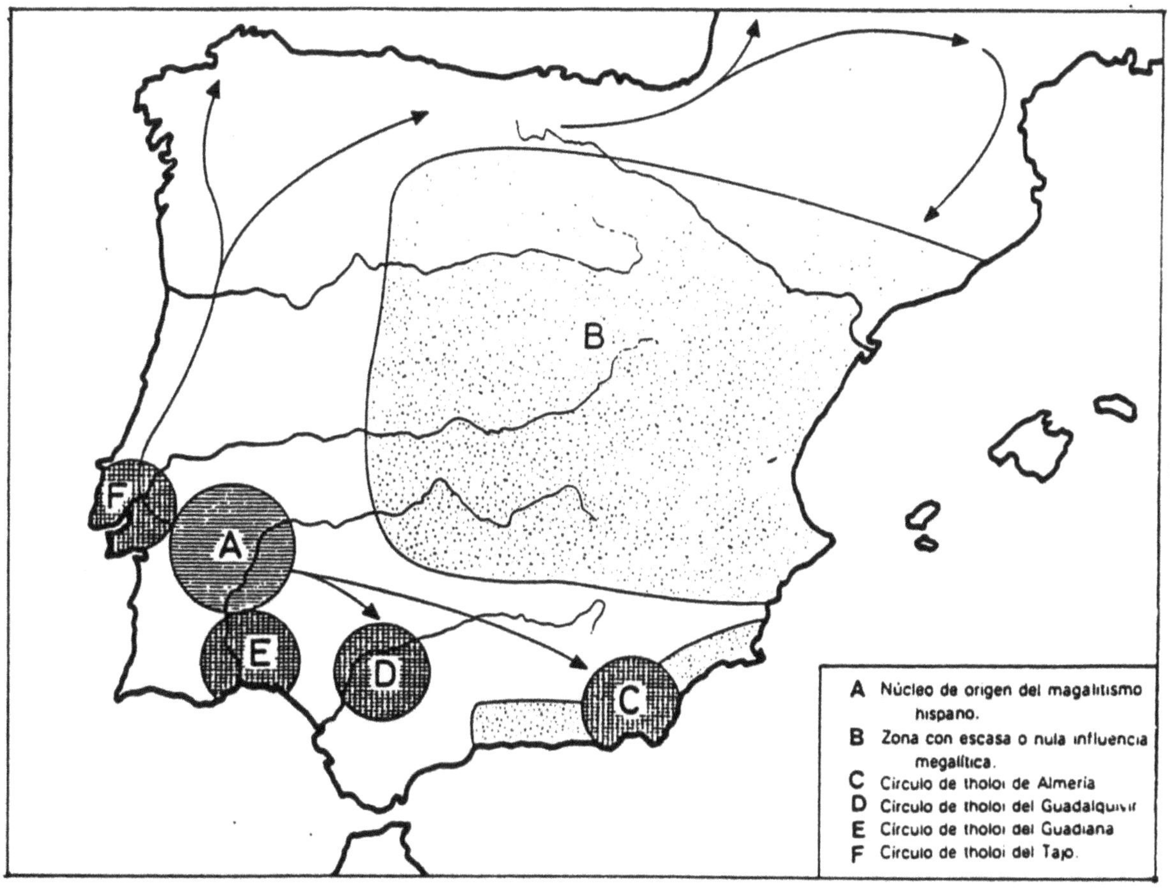

Fig 7.2 MEGALITHS IN SPAIN
Jorda Cerda, F.; Pellicer, C. M.; Acosta, P., Almagro-Gorbea, M., 1989, 240

The first use of Megalithic burial structures in the Iberian Peninsula appears around (4500 BC - 3500 BC) depending on the region and archaeological dating methods used. At present it is thought the megalithic structures first appeared in the Portuguese South East region centred around the present day provinces of Ribatejo and Upper Alentejo. The use of Megalithic burial structures then spread almost immediately to the South East and West Atlantic coastal regions of Iberia (Jorda Cerda, 1989, 240).

Megalithic tomb structures are thought to have spread into the Mid-Central Iberia region around 3200 BC - 2800 BC (Fig 7.2) from the West or South Atlantic Coastlands via the Guadadina, Tagus and Duero Rivers (Savory, 1968, 107-9). Megalithic tomb structures are represented in Mid-Iberia with a considerable concentration in the region of Valencia de Alcantara (Province of Badajoz) close to or on the present day Portuguese/Spainish border (forty three dolmens surveyed to date) and seventy five dolmens in small group concentrations within the province of Salamanca (Delibes y Santonja, 1986, 18-19).

The Eastern side of the present day Autonomous Community of Extremadura represents the eastern boundary of megalithic tomb influence in the Southern Meseta geographical region.

The start of the Chalcolithic period in Mid-Iberia is set at about 3200 BC - 2800 BC. In the mid-Iberian region, the Chalcolthic period arrived later than on the West and South Coastlands and is represented by archaeological work done in the 1950's and 1960's along the banks of the Guadiana river in the region of Merida 2000 BC -1800 BC (excavations by Dr. Enriquez Navascues, Merida). Most of the present day pre-Bronze Age archaeology excavations being done by the University of Extremadura and other Spanish institutions within the Extremadura region are centred on Chalcolithic sites (Navascues and Diaz, 1991, 11-158).

The appearance and use of megalithic tombs in the Southern and Western Coastlands regions is generally accompanied by the first use of copper metal for tools and weapons in the Iberian Peninsula (Savory, 1968, 138). The beginning of the III millennium BC is the start of the Copper Age in Mid-Central Iberia, which lasts until the start of the II Millennium BC.

The primary periods of antiquity covered by this archaeometallurgical tin survey include: the Copper Age, the Bronze Age and the Roman period to 4th century AD.. Iron started appearing in Iberia with the influx of Phoenician and Greek traders around 800 BC. By the beginning of the Roman period (150 BC) in Mid-Central Iberia, iron had become firmly established as the dominant metal.

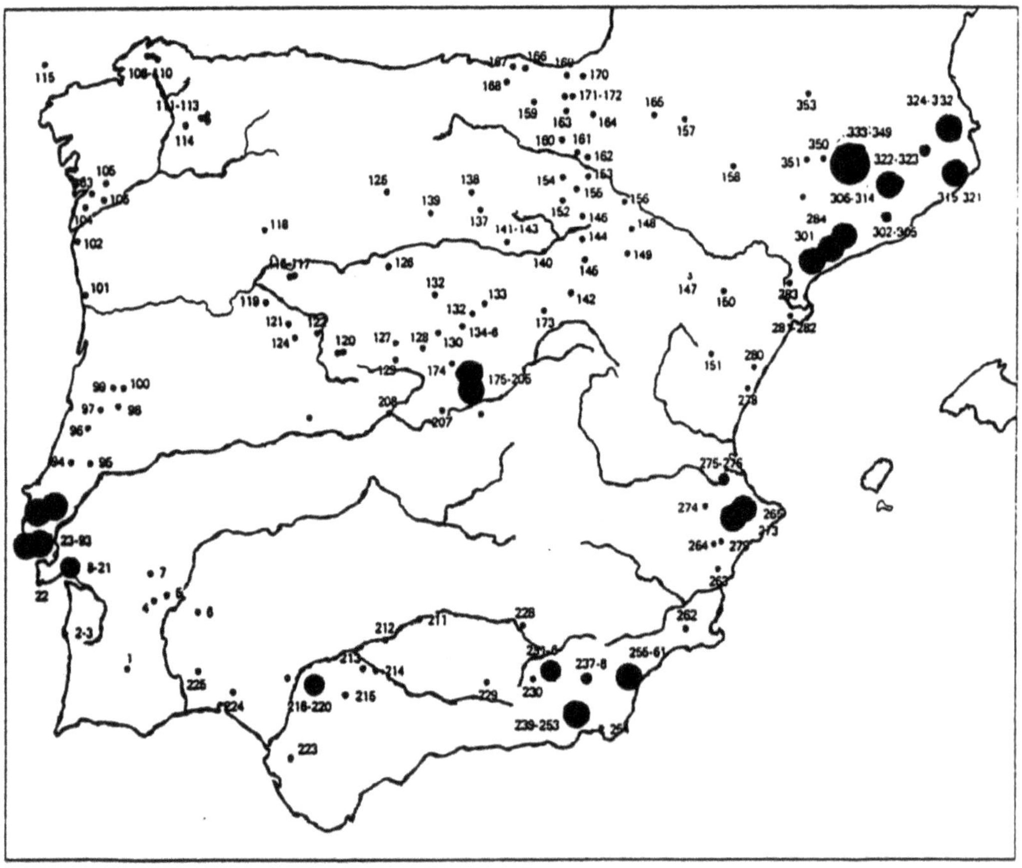

Fig 7.3 DISTRIBUTION OF BEAKER SITES IN IBERIA
Harrison, R.J., 1977, 100

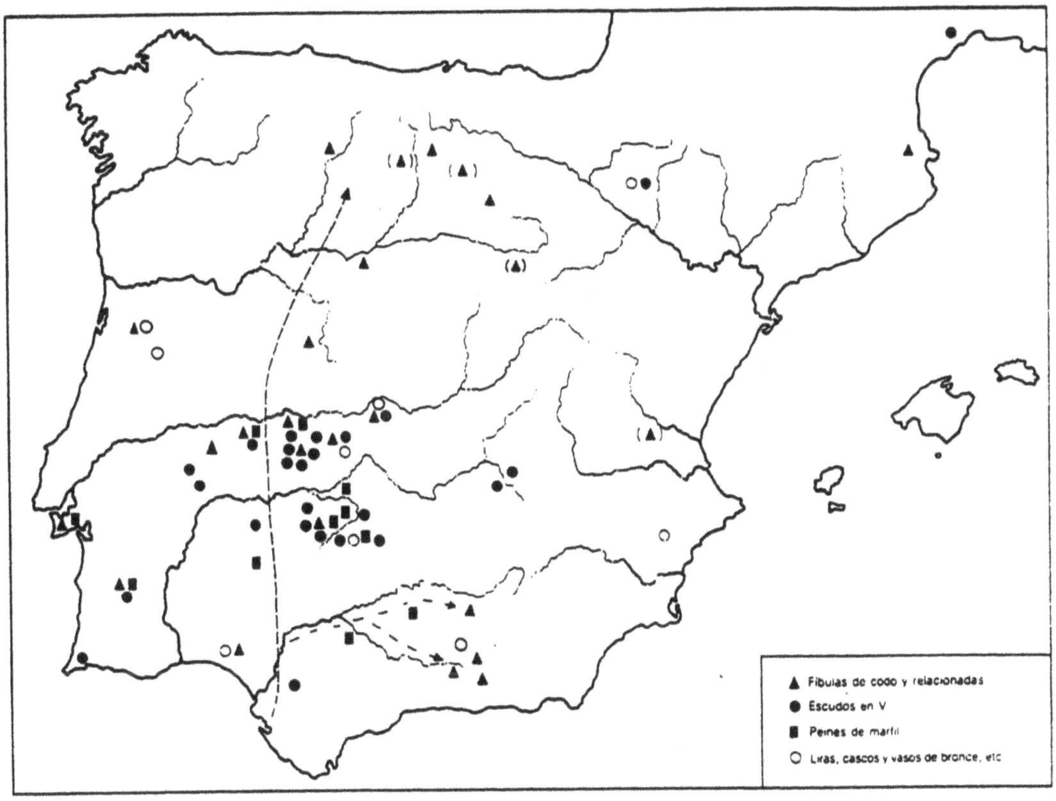

Fig 7.4 Dispersion of Proto-Oriental Elements most significant in the Iberian Peninsula
Jorda Cerda, F.; Pellicer, C.M.; Acosta Martinez, P.; Almagro-Gorbea, M., 1989, 432

The Bronze age in the Mid Central Iberian area starts from about 2000 BC. and last till approximately 700 BC. The region of Spain and Portugal in which I did my survey work is a region within the Iberian Peninsula where until quite recently very little archaeological work, neither survey or excavation have been done on the Copper/Arsenical Copper/Bronze Age sites.

It is in the late III millennium and early half of the II millennium that the Beaker material complex starts appearing within the archaeological record in certain regions in the Iberian Peninsula (Harrison, 1980, 126-28).

Within the Mid-Western Central region the Beaker dates are approximately (1900 BC - 1600 BC). To date the archaeological finds of Beaker pottery and metal artefacts are few within the region I surveyed (Fig 7.3). The province of Salamanca has five sites and the province of Caceres one site with Beaker material present (Harrison, 1977, 100).

The largest concentrations of Beaker sites during the late III and early II millennium were located in the Northern Meseta/Central Sierra regions (the Campozuelos Beaker Complex) and the in the Tagus Estuary (the Maritime Beaker Complex) and adjoining West Atlantic Coastlands regions (Harrison, 1977, 100-241).

The lack of Beaker material within the Southern Meseta geographical region could be due to the small number of archaeological surveys and excavations that have been undertaken in this region. Future excavations within the Southern Meseta region should provide more Beaker material evidence.

It would seem logical that the Tagus and Duero river valleys would have been the route that allowed the Beaker complex (Atlantic and Campozuelos) to share trade or to have a influence upon existing cultures within the Mid-Central Iberia region between the Duero and Tagus Rivers.

It is thought that most of the cultural influence and any trade that took place in prehistoric times (Late Mesolithic, Neolithic, Chalcolithic) in the Mid-Central Iberia regions came from the Atlantic Coastland region of the Iberian peninsula.

By the late II millennium and early I millennium (1200 BC - 700 BC - the Late Bronze Age) the major culture influence to the Mid-Central Iberian region was coming from the Tatressian Culture whose centre was located in the South Western Atlantic and Mediterranean Coastland regions of the peninsula (Fig 7.4), or in what is today the western half of the modern region of Andalusia (Jorda Creda, 1989, 527).

Tartessian culture and trade influence extended into Mid-Central Iberia as far as the present day town of Caceres (Jorda Cerda, 1989, 432-33). The most widely accepted theory for the rise of Tartessos and its wealth is considered to be the Tartessian control of the mining and trade of metals tin, copper, silver, gold lead and iron (Arribas, 1963, 49-50).

It was during the Tartessian period that wheel thrown ceramics, first introduced into Iberia by the Phoenicians (FIG 7.5) reached the Central-Mid Iberian Region.(700 BC - 575 BC). It then took until the Roman conquest of the Lusitanian region (300 BC - 133 BC) for the introduction of wheel made pottery to move further North into the Duero River upper catchment area (Jorda Creda, 1989, 444).

The Tartessian Culture of South Western Iberia was strongly influenced by the Phoenician coastal trade settlements on the South Mediterranean and Atlantic coasts (Aubet, 1987, 218-47). According to historical sources the influence of the Tartessian culture and trade was replaced by the Carthaginians around 500 BC (Aubet, 1987, 273-76).

At the same time the Turdetania culture (either replaced or followed from) the Tartessian culture within the South-West region. The Turdetania influence and trade extended North to areas of Mid-Central Iberia region as had the Tartessian. In this region the Turdetania and Lusitanian cultures existed together.

The Lusitanian culture was centred between the middle region of the Tagus and Duero Rivers and influenced from the East Atlantic Coastlands.

During the period 550 BC to 206 BC, the Carthaginians took direct control of some of the Iberian cultural areas on the East and Southern Mediterranean Coastlands. The Carthaginians at the same time tried to extend their influence into the Meseta Regions both North and South of Central Spain (Galan, 1991, 391-98).

The Romans gained complete military and political control of all the Carthigian regions in the Iberian Peninsula after a war with Carthage lasting from (218 BC - 206 BC) (Keay, 1988, 27-9).

By 200 BC the Lusitania culture had became dominant in the Central Mid Iberian region, replacing the Turdetania culture. The Lusitanian region included most of modern Portugal South of the Duero and the Spanish provinces of Salamanca and Caceres and about fifty per cent of the province Badajoz.

The Lusitanians had a series of wars with the Romans from 155 BC - 139 BC in which Rome gained complete political control of all Lusitanian territory (Keay, 1988, 33-5).

The Lusitania region after the Roman Celtiberian Wars of 155 BC - 133 BC became part of the Roman administration district of Lusitania which was in turn subdivided into three judicial districts: Conventus Palensis, Conventus Scalabitanus and Conventus Emeritensis (Edmondson, 1987, 12). The Lusitanian judicial district of Conventus Emeritensis roughly corresponds geographically to the region in which I did my tin survey work.

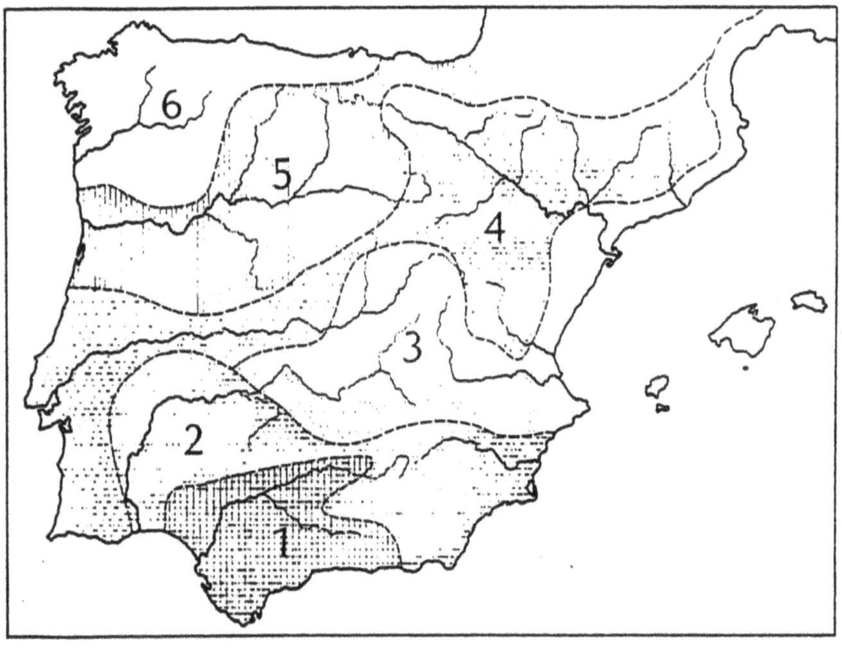

Mapa de la difusión de la cerámica a torno en la Península Ibérica.

1. Antes del 700 a. C.; 2. 700-575 a. C.; 3. 575-425 a. C.; 4. 425-300 a. C.; 5. 300-133 a. C.; 6. Después del 133 a. C.

Fig 7.5 Map of the Difussion of Wheel Made Pottery in the Iberian Peninsula
Jorda Cerda, F.; Pellicer, C.M.; Acosta Martinez, P.; Almagro-Gorbea, M., 1989, 444

The following chronology has been provided to me by Alonso Rodriguez Diaz and Ignacio Pavon Soldevila.

During my survey within Extremadura I have been advised and assisted by Alonso Diaz and Ignacio Pavon Soldevila of Caceres University, as to the dating of any surface pottery which I was able to collect.

Alonso Diaz is a lecturer in prehistory at the University of Extremadura and Ignacio Pavon is a PhD student at the same University. Ignacio Pavon's PhD work has been based on the first survey of Late Bronze Age sites within Extremadura. The majority of Ignacio Pavon's work has been done within the province of Caceres; his doctorate work has included excavations and stratigraphy of two Bronze Age sites.

The following dates from Extremadura are based on pottery and not radiocarbon charcoal dating.

EDAD DEL COBRE (COPPER AGE) III Millennium (3000-2000) BC
only copper is appearing in Extremadura at this time

EDAD DEL BRONZE (BRONZE AGE) II Millennium (2000-1000) BC
the Edad del Bronze is further divided into the following three
sub classifications

I (Bronze Inicial) (2000-1500) BC
the beginning of the bronze period
bronze does not appear in the Mid-Iberia Extremadura region
this period has only copper and some arsenical copper

II (Bronze Pleno) (1500-1200) BC
this is called the full Bronze period
bronze does not appear in this period either;
copper and arsenical copper appear

III (Bronze Tarado) (1200-900) BC
this is called the late Bronze period
bronze does start to appear
but this is chiefly a period of arsenical copper

BRONZE FINAL (900-700) BC
this is called the final Bronze period
this is the true bronze period in Extremadura
Bronze is now appearing in all archaeological contexts
and arsenical copper has disappeared

In Portugal I have been helped with the surface pottery identification from the Portuguese survey sites by Paulo Felix and Luis Oosterbeek of the Archaeology Department, Escola Superior de Tecnologia e Gesto Tomar, Portugal.

I have also been assisted in pottery identification and local archaeological chronology by Raquel Vilaca of the Instituto de Arqueologia, Faculdade de Letras, Coimbra, University.

Raquel Vilaca has also been finishing her PhD work in the last few years. Her PhD work has also been based on the surveying of settlement sites and the excavations of four Late Bronze Age settlements within the Portuguese province of Beira Baxia.

Raquel Vilaca has given me calibrated radiocarbon dates for the Late Bronze Age sites (Bronze Final) she has excavated as 1200 BC - 900 BC.

I was also assisted in Portugal with pottery identification for the dating of the surface slag which I found at Torre Romana Centumcellas. Maria Helena Simoes Frade (The Regional Director do IPPAR de Coimbra) director of the excavations at the Torre Romana Centumcellas site has confirmed that the collected survey surface tin slag from this site is Roman. Similiar slag fragments appeared in her excavated areas.

Most of the small amount of surface pottery that I was able to find on my survey site has been dated to the Late Bronze Age period (Bronze Final). There are two exceptions to this, the site of Berrocal in Spain which has been dated to the Neolithic-Chalcolithic around 2000 BC and the site of Torre Romana Centumcellas in Portugal which has been dated to approximately 1st to 4th Century AD.

Ignacio Pavon Soldevia and Raquel Vilaca give me their metal artefacts and any supposed casting or crucible fragments from their survey and excavation work. The metal and crucible fragments were analysed in London at the Institute of Archaeology. This analytical work has now been finished and the results seem to support the metallurgical sequence of the chronology presented on page 70. In that the metal artefacts of copper (Copper Age), arsenical copper (Bronze Age) and bronze (Final Bronze Age) seem generally to stay within their accepted chronologicial periods.

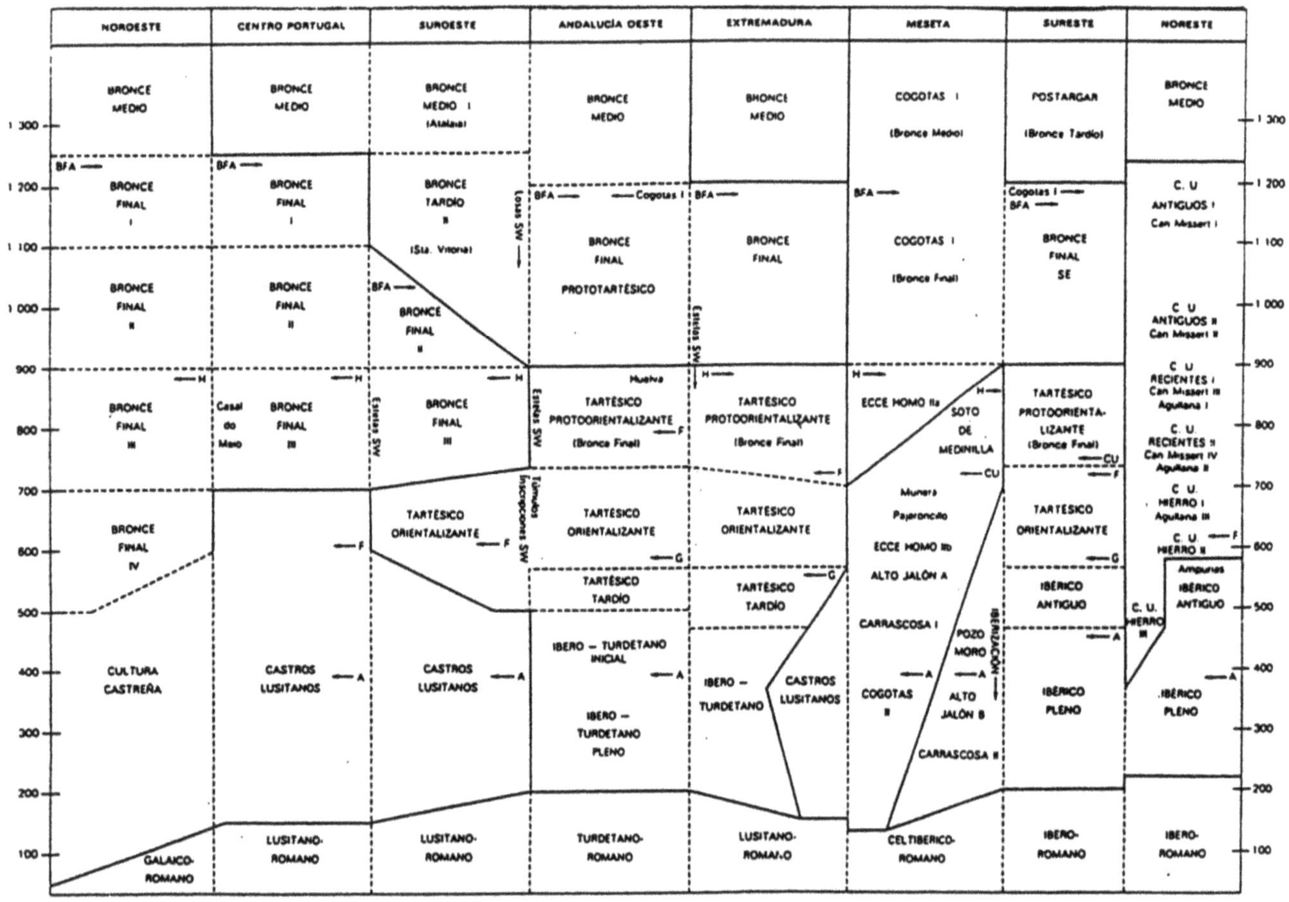

Cuadro de la Península Ibérica durante el Bronce Final y la Edad del Hierro entre el 1300 y el 0 a. C.

BFA = Bronce Final Atlántico; H = Bronce Final de Huelva; CU = Campos de Urnas; F = influjo fenicio; G = influjo 'greco-focense'; A = horizonte de importaciones áticas

Fig 7.6 Table of the Iberian Peninsula during the Bronze Final and the Iron Age 1300 BC to 0 AD
Jorda Cerda, F.; Pellicer, C. M.; Acosta Martinez, P.; Almagro-Gorbea, M., 1989, 343

SITES SURVEYED

SPAIN

PORTUGAL

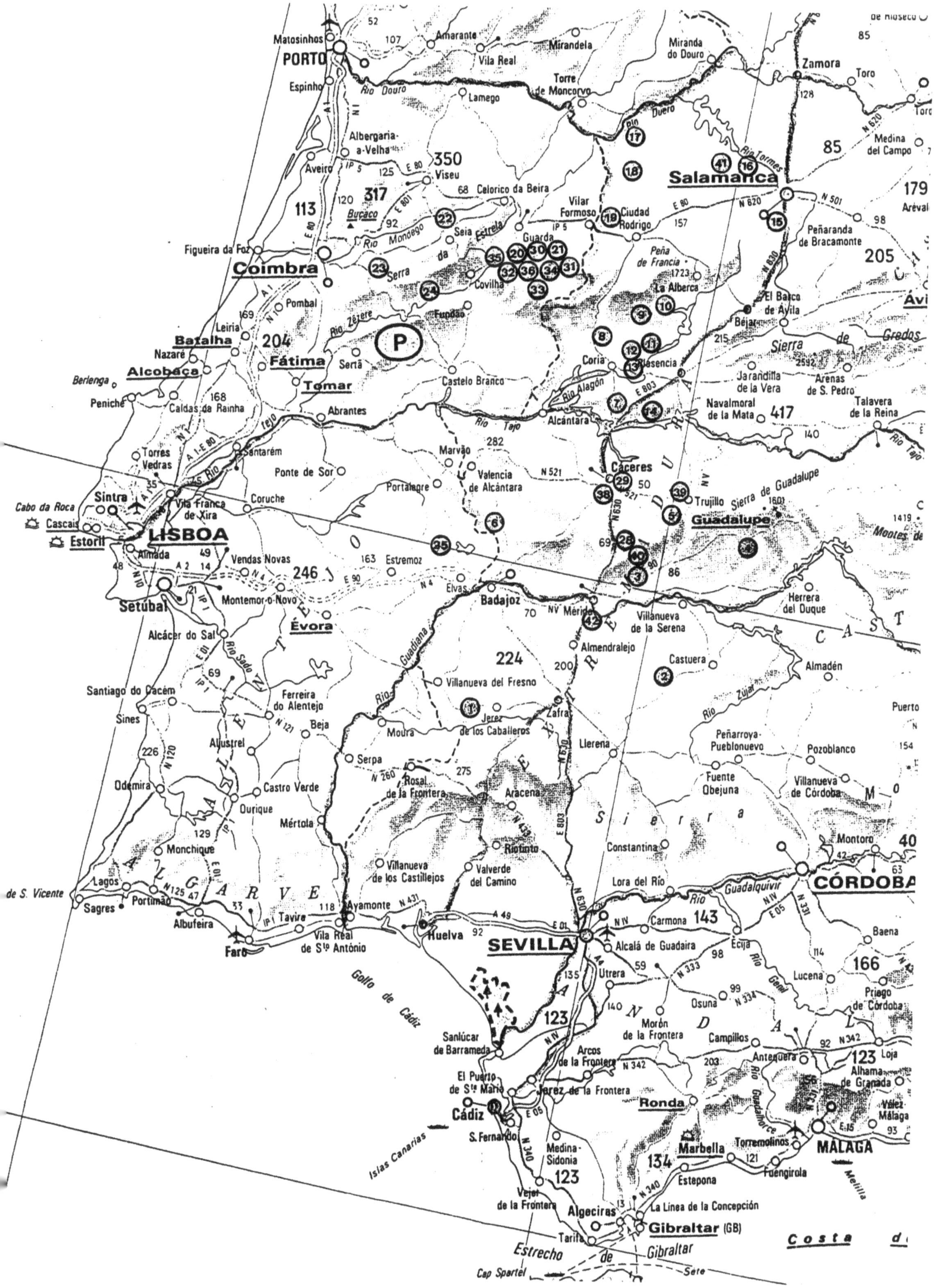

Fig 8.1 SURVEY SIE LOCATION MAP
EUROPA, MICHELIN 970, 1:300.000 (1CM:30 KM), 1995

VISIT NO 1 **MINA MARI JULI Y MINA VIRGIN DE GARCIA**

PROVINCE: BADAJOZ

MUNICIPALITY: Oliva de la Frontera and Jerez de los Caballeros

CO-ORDINATES: WEST: 06° 56' 00" - NORTH: 38° 19' 05"
Provinical Map BADAJOZ 1:200.000 1989 2a ed.

Photo 1 North of Zahinos - Oliva de la Frontera Small Open Cut Mine

This mine site was listed in one of the mine reference books which I was using (Dunning et al., 1989, 128) as a tin-wolframite mine. Being the most southern mine on my list of possible ancient tin mining sites Oliva de la Frontera became the first site that I visited. When arriving in the town of Oliva de la Frontera I asked various local people in the town if they had heard of the two specific mines that I was looking for. No one knew where they were located, nor had they heard of them. No one knew of any tin mineral ever being mined nearby in ancient or modern times.

I spent 5-6 hours driving and walking south, east and north of Oliva de la Frontera trying to locate the mines but had no luck; neither did I have any luck finding local people who knew where the mines might be located. So I decided that this first visit site was not worth any more of my time.

I later found out in the summer of 1994, after receiving the latest edition of La Minera En Extremadura (Molero, 1993, 113-15), that the mines by the town of Oliva de la Frontera were only wolframite and bismuth mines and mined only in the modern era (1930).

The above publication also stated that Roman copper smelting evidence existed in the area of Oliva de la Frontera. The Romans had also undertaken alluvial working for gold along the Rio Ardila which runs in a South/West to North/West direction about three kilometres to the South and East of Oliva de la Frontera.

The area of Oliva de la Fronteria yielded no evidence of cassiterite mineralisation or any ancient mine sites. Therefore the Santa Eulalia Mine, in the Portuguese Province of Alentejo, (Visit No 25, page 100 was the most southerly cassiterite mineral site that I visited.

VISIT NO 2 **GROUPE MINA SAN NICHOLAS**

PROVINCE: BADAJOZ

MUNICIPALITY: Valle de la Serena

CO-ORDINATES: WEST: 05° 52' 20" - NORTH: 38° 40' 15"
Provinical Map BADAJOZ 1:200.000 1989 2a ed

Photo 2 General View from East - Mina San Nicholas

This site proved quite easy to find. The production of Mina San Nicholas in the modern period was mostly wolframite along with some cassiterite and some bismuth in a ratio of 3/1/5. In the modern period the mine was worked 1912 to 1945 and 1986 to 1990. It was a gallery/shaft mining system consisting of approximately seven galleries running in the general direction of S/W to N/E with a average working depth of approximately 60 meters with cassiterite appearing in only two of the galleries along with bismuth La Minera En Extremadura (Molero,1993, 109-11). In all references of the Mina San Nicholas there is no mention of archaeometallurgical remains or ancient pottery.

I spent almost a full day at the Mina San Nicholas. The only thing that I collected during this time was some small supposed mineral samples. No archaeometallurgical mining or smelting remains or pottery were discovered. I did not go into any of the mining galleries because they were in quite a dangerous state.

It is possible that in the early stage of existence that the Mina San Nicholas would have had surface veins of cassiterate which have since been worked out. If these areas were precisely located there might be some chance of finding early mining or metallurgical remains nearby.

The Mediterranean bush and dwarf shrub vegetation growth at this mine site made it very difficult to do a very thorough walking field survey in the short time that I was there. I think that this site needs more intensive surveying of what is quite a large area. Perhaps in the future, time permitting I will pay another visit to the Mina San Nicholas as I still feel that there is some possibility of this mine having been worked for vein cassiterite in antiquity.

VISIT NO 3 **MINA LA PARRILLA**

PROVINCE: CACERES

MUNICIPALITY: Almoharin

CO-ORDINATES: WEST: 06° 07' 10"- NORTH: 39° 05' 00"
Provinical Map CACERES 1:200.000 1989 2a ed

Photo 3 View of Waste Heaps La Parilla

The La Parrilla mine is a large open cast modern mine. The mine was in operation from approximately 1930 to 1987. During this period La Parrilla was the largest producer of the mineral scheelite in Spain. During the same period small amounts of cassiterite were also mined at La Parrilla, but only in ratio of 12 to 1; with scheelite production being the larger (La Minera En Extremadura) (Molero, 1993, 103-6). The cassiterite and scheelite ores were located on the surface and down to a depth of 50 meters in very fine veins.

The La Piralla mine site due to its modern opencast working methods and size of operations presented a very confusing and difficult mine area in which to do an archaeometallurgical survey. I spent about 6 hours driving and walking around the mine, but was not able to find any landscape area within the La Piralla mine that might prove rewarding for a more intense field walking survey.

Though there is reference to ancient workings in La Minera En Extremadura, I would think that any sign of this evidence would have been removed when the large modern open cast mine was brought into production.

I found no evidence of ancient mining, smelting or pottery. The La Pirilla mine site due to its extensive size will need a great deal more time spent at it, if any ancient archaeolometallurgical evidence is to be found there. Never the less, more intensive surveying seems worthwhile. At the initial stage of my Iberian survey work project I did not have the time avaivable that the La Parrilla site required.

VISIT NO 4 **LA MINA EL CERRO DE SAN CRISTOBAL**

MUNICIPALITY: LOGROSAN

CO-ORDINATES: WEST: 1° 50' 50"- NORTH: 39° 19' 38"
Hoja VALDECABALLEROS 1:50.000 1a edicion 1953
General Instituto Geografico y Catastral

Photo 4 View of El Cerro De San Cristobal from 25 kilometers distance, looking East

When I first arrived at Logrosan municipality in the summer of 1992 I did not have a great deal of information to go on at that time, except that the Logrosan area was listed as containing a Tin-Wolframite Mine.

The Logrosan area has been the site of a very important Spanish phosphate mining and production operation from (1775-1921). The phosphate vein areas at Logrosan were approximately 300 meters below the summit of El Cerro de San Cristobal in the West Filon Mingote, North Filon Ginjal and East Filon Constanza.

The three main mined phosphate veins are located primarily in the surrounding slate area although two of the veins (Ginjal and Constanza) are present for short distances in the metamorphic aureole of Logrosan. The three phosphate veins run basically in a N30E direction (Barrena and Isidoro, 1987, 73-4).

In the period 1917-1921 the phosphate mines at Logrosan produced 90,000 tons of phosphate, which was one half of the national production for Spain at that time (Molero, 1993, 106).

El Cerro de San Cristobal itself is a solitary granite batholith running in a SW-NE direction, approximately two and one half kilometres in length East to West and one kilometre in width North to South. El Cerro de San Cristobal is completely encircled by a well marked metamorphic aureole (Fig 8.2).

El Cerro de San Cristobal is completely surrounded by a slates and sedimentary structure rocks, the nearest large granite region is approximately 20 kilometres to the west appearing by the town of Zorita. Six kilometres to the North of Logrosan is a range of high mountains called the Sierras de Guadalupe which average well above 1000 meters.

INVESTIGACION LOGROSAN (CACERES), 1980

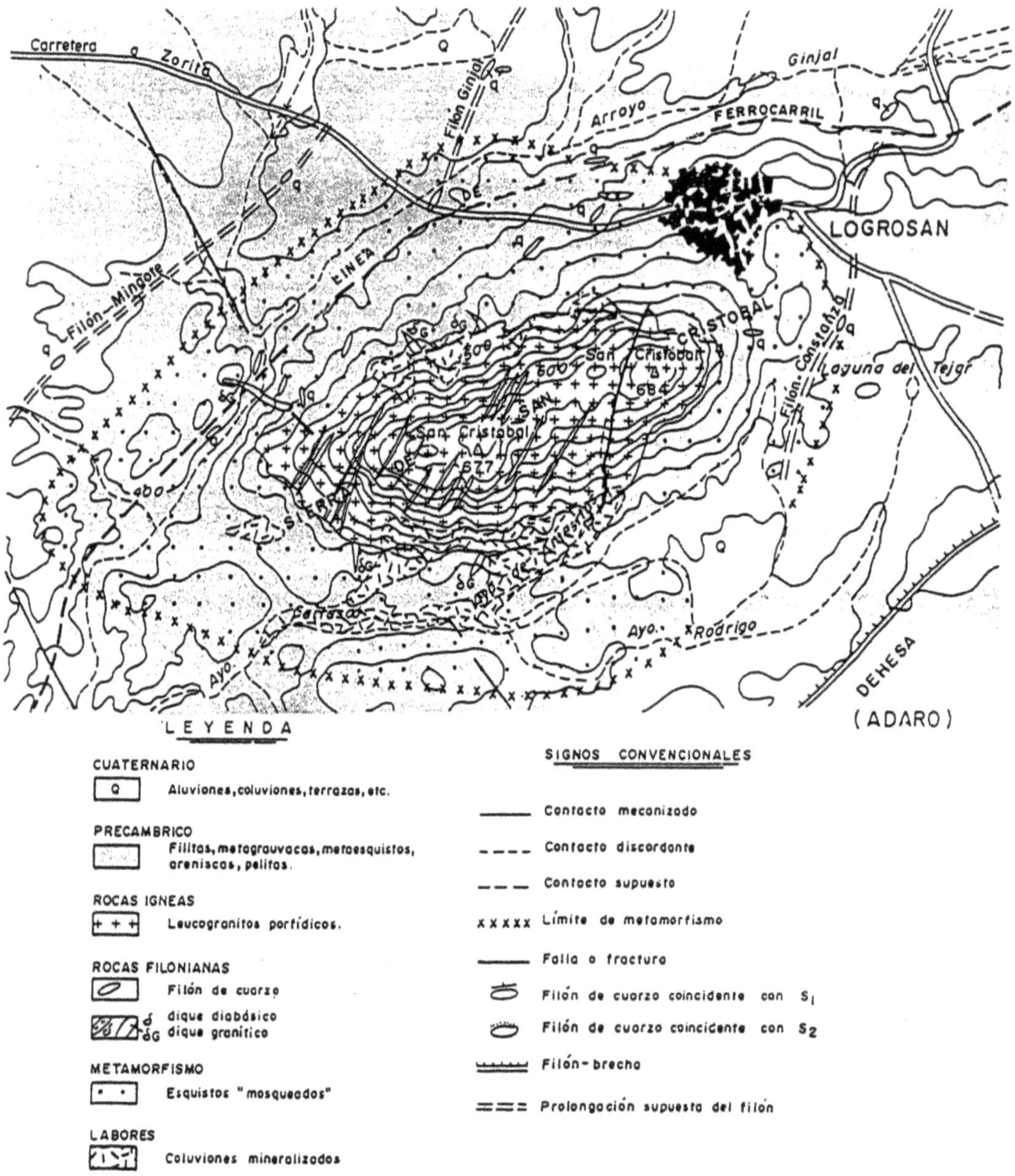

Fig 8.2 INVESTIGATION LOGROSAN (CACERES), 1980
Barrena, M.A. and Isidro, P.D., 1987, 74

Six kilometres to the south of Logrosan is a large river valley which contains the Rio Ruecas, Rio Cubilar and the Rio Gargaligas all tributaries which flow into the Rio Guadiana, which is located in the same large river valley. This river valley consists of alluvial terraces consisting of gravels, sands and conglomerates.

The highest point of the Cerro is approximately 670 meters above sea level and 200 meters above the present level of the main town square of Logrosan, (population approximately 4000) on the lower North East slope of El Cerro de San Cristobal.

The mountain of El Cerro de San Cristobal stands out very acutely in a flat low rolling hill region which runs from the Sierra de Guadalupe to the Guadiana River Valley. The view from the top of the El Cerro de San Cristobal in all directions can only be described as majestic.

The Logrosan granite is classified as a leucogranite porphyry with two micas and formed from phenocryst of quartz and small amounts of feldspar, biotite, apatite, and rutile with some iron oxide present. There are three areas on the Northern slopes and one area on the Western slope consisting of kaolinized granite (Sos Baynat, 1967, 329-30).

Within the area of Logrosan both in the slate and granite are numerous quartz veins in some cases up to a meter wide and running for some kilometres. Not all of the quartz veins are mineralised (Barrena and Isidro, 1987, 73).

The cassiterite mineral (Sample No 103, page 135) veins which run in a N30E/50SW direction are found only within the granite of El Cerro de San Cristobal. Present also in the granite in varying amounts are stannite ($CuFeSnS_4$) and arsenopyrite (FeAsS) (Sample No 102, page 135) accompanied by smaller amounts of chalcopyrite ($CuFeS_2$), pyrite (FeS_2) and sphalerite (ZnS) (Barrena and Isidro, 1987, 74). Most of the cassiterite veins are accompanied by quartz and normally by tourmaline (Sos Baynat, 1967, 334-61).

In 1949 local people and traveling miners searching for gold in the Logrosan area led to the rediscovery that the mineral cassiterite was present in large amounts on the mountain of El Cerro de San Cristobal. After 1949 a very chaotic system of exploration and mining of the cassiterite veins, elluvial and alluvial deposits took place on most areas of El Cerro de San Cristobal.

I did not so much discover this archaeometallurgical mining site as rediscover a site which had in the recent past (1950's-1960's) been looked at quite extensively geologically/geographically and had been noted as a site with a diversity (pottery, flint, stone) of archaeological surface remains (Molero, 1993, 106).

Vicente Sos Baynat (1902-1993), a noted Spanish geologist living and working in the province of Extremadura, spent a lot of time looking at and writing about the geology and geography of the Sierra de San Cristobal during the 1950's and producing a very detailed geological report on del Cerro de San Cristobal and its mineralisation (Sos Baynat, 1965, 295-447).

My first survey visit to El Cerro de San Cristobal was most fortunate in its timing, in that it took place shortly after a small area of the north western top of the Cerro, just to the north west of the 677 survey marker, had been cleared of all vegetation by a brush fire. This seems to have happened only a week or two before my arrival.

The fire cleared area proved very rich in visible archaeological remains: pottery (Drawing No 3 & Drawing No 4, pages 58, 59), pages, stone loom weights, stone hammers and crushing stones fragments and what seemed like the remains of stone wall foundations of what had been round structures.

Further survey walking on the western top of the Cerro de San Cristobal led to finding the fragment of a possible grooved stone mining hammer (Drawing No 1, page 57). Most of the western top area of the Cerro was covered with ancient pottery.

I also encountered a few areas on the South-East area of the Cerro which were extensively covered with crushed quartz heaps and littered with crushing-grinding stones and accompanied by extensive pottery fragments (Photo 5).

The exact date of the majority of the surface collected pottery (Late Bronze Age) would not be known till I had the pottery examined at University of Extremadura, Caceres, at the end of my first survey season.

A large percentage of the stone hammers found on the surface during my surveys at El Cerro de San Cristobal came from rocks that had been formed in the metamorphic aureole surrounding San Cristobal (Drawing No's 5, 6, 7, 8, 9, 10, pages 60-65). This rock is a extremely hard and durable metamorphosed slate (Dr Brinley Roberts, personal communication 1995). A large percentage of the aureole hammers had also been shaped and smoothed by water action in seasonal river beds which flow on the North and West sides of El Cerro de San Cristobal.

I returned to El Cerro De San Cristobal twice in 1993 and twice in 1994. Each new visit provides more evidence that the tin mine site on the San Cristobal mountain will be one of the most interesting and rewarding of my survey work.

In the first 1994 visit I took along a two person site planning team which was to start and hopefully finish a site plan of 1:500 of the Late Bronze age mining and settlement site which seemed to centre on the Western side of El Cerro De San Cristobal. (A standard Hilger Watts ST156 Theodolite was used during the first year's survey).

It was not possible to finish the site plan of the Western top area of El Cerro de San Cristobal in 1994, due to the large area that the Late Bronze Age settlement site covered and the complexity that the geographical setting the site presented to the planning team.

Typical Extremadura April-May weather provided some days of extreme cold and rain. During the cold wet weather no planning work could be done on the mountain top.

In 1995 I again returned in April-May to the site with the same planning team. The main aim of the 1995 season was to try to finish the basic planning surveys of the Cerro. Due to good weather and the use of modern electronic surveying equipment (Nikon DMT 20 Total Station) the basic ground and contours survey maps were finished in the 1995 season (Plan 1 and Plan 2).

During the 1994 and 1995 seasons at Logrosan more surface evidence was noted and collected which pointed most strongly to El Cerro De San Cristobal being what I would now definitely consider to be a walled Late Bronze Age Settlement site on the top of a granite mountain which at one time must have contained quite a large amount of veined and elluvial tin deposits.

The next stage at Logrosan in 1996 will be to map exactly the location of the former quartz bearing tin veins and any known mine galleries both ancient and modern.

It is also planned in the near future (if funding is available) to do a season of small excavation trenches with the University de Extremadura to try and ascertain a proper archaeological sequence for El Cerro de San Cristobal and hopefully to find some stratified charcoal which will give some definite dates to this very interesting site.

La Mina El Cerro de San Cristobal has with time and further survey work turned out to be one of the more important tin mining sites that I have yet run across in my Iberian Tin survey project.

Photo 5 Broken Quartz and Pottery Fragment North East area of El Cerro de San Criistobal

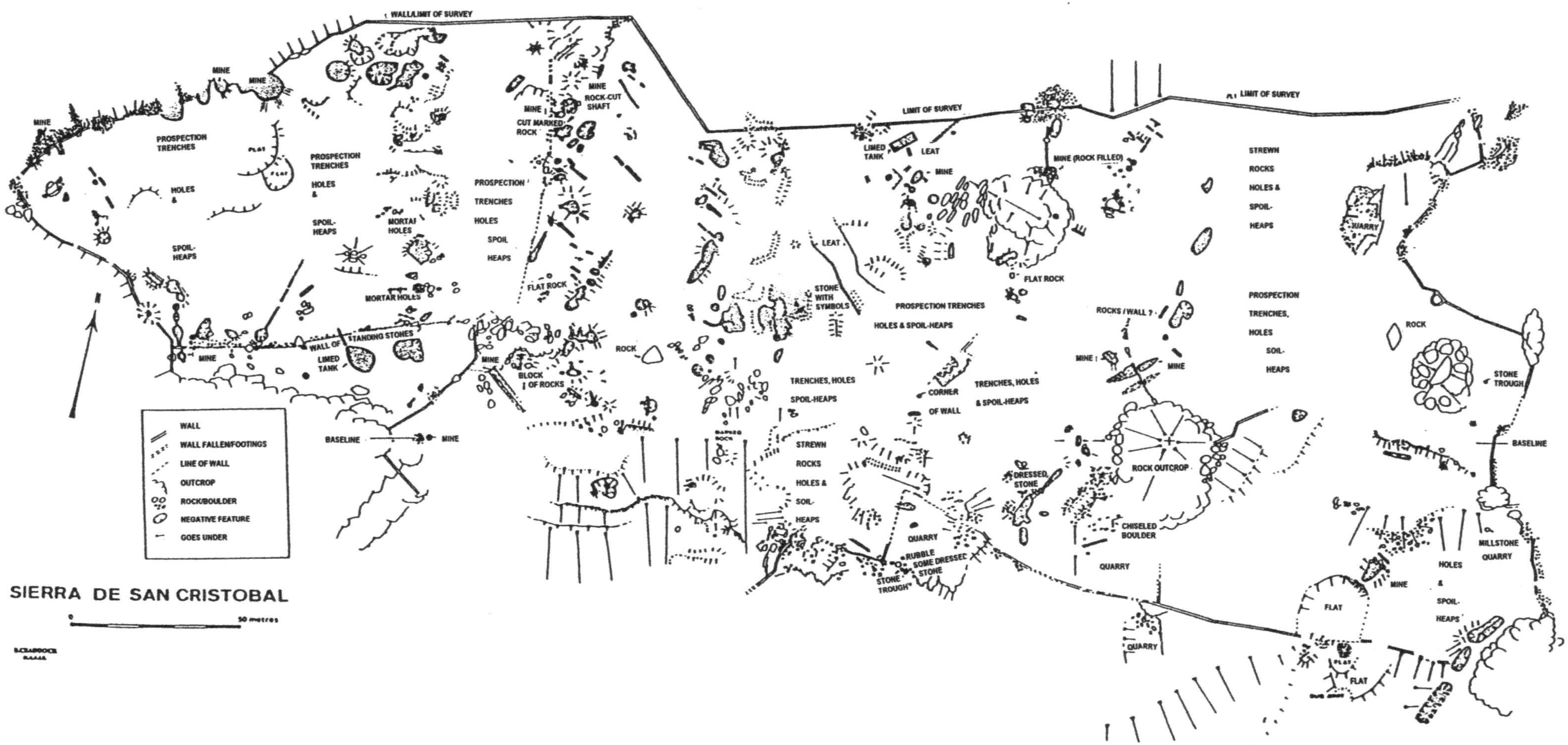

SIERRA DE SAN CRISTOBAL

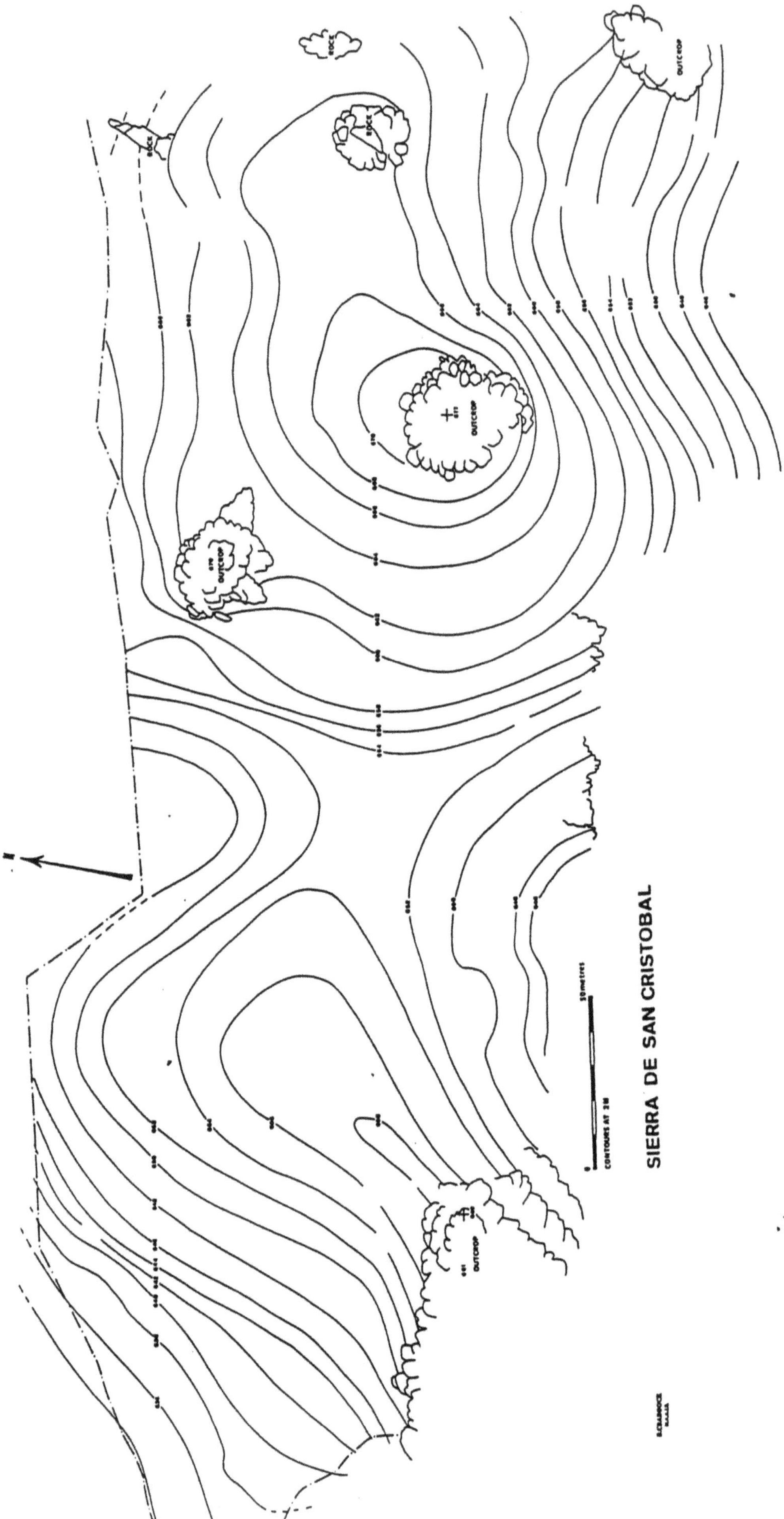
SIERRA DE SAN CRISTOBAL
OUTCROP
ROCK
CONTOURS AT 2M

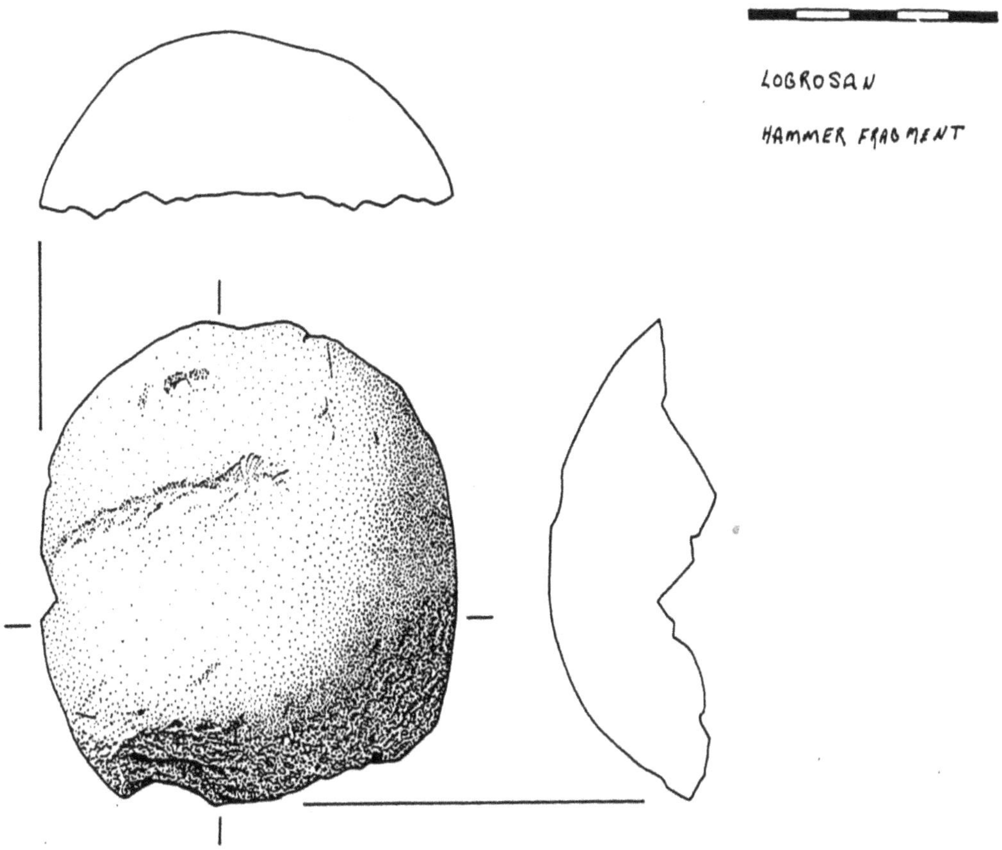

Drawing No 1 fragment stone hammer, San Cristobal, scale units 1cm

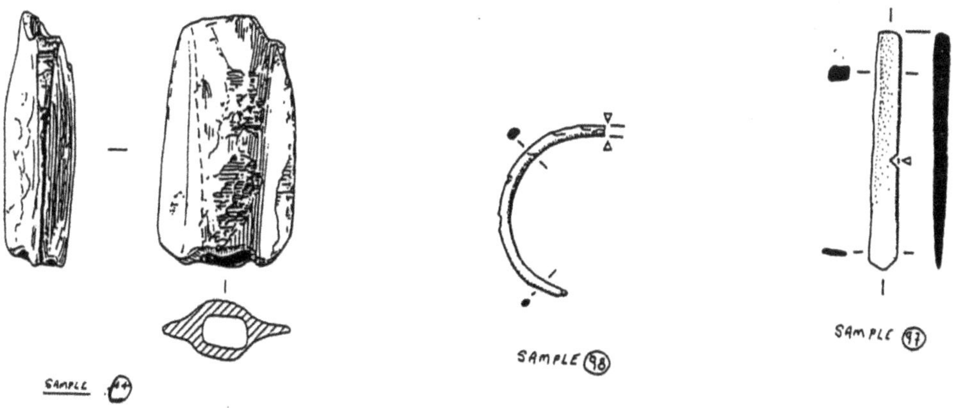

Drawing No 2 fragments three metal objects, San Cristobal, scale 1/1

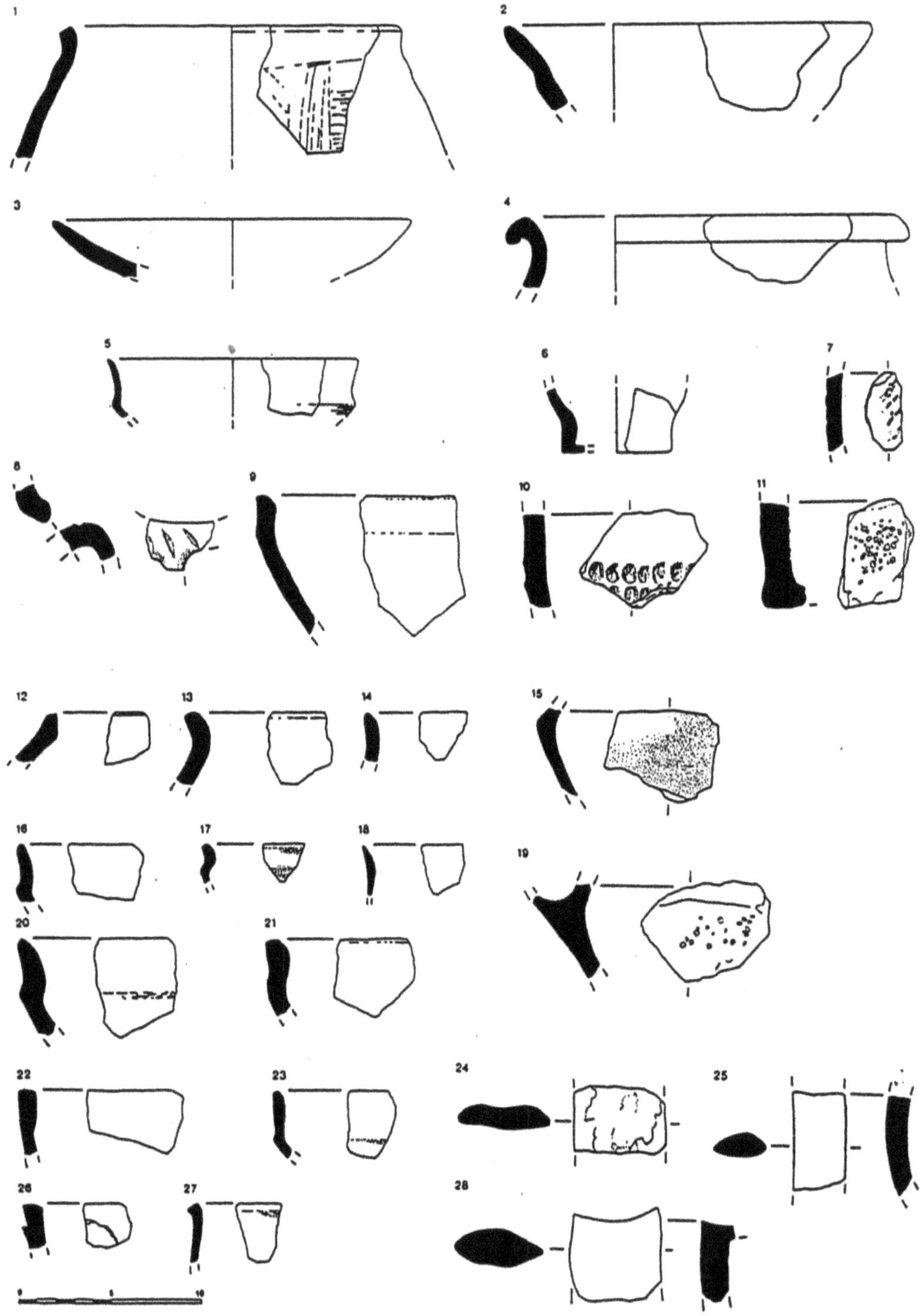

Drawing No 3 survey pottery, San Cristobal, scale units 1cm

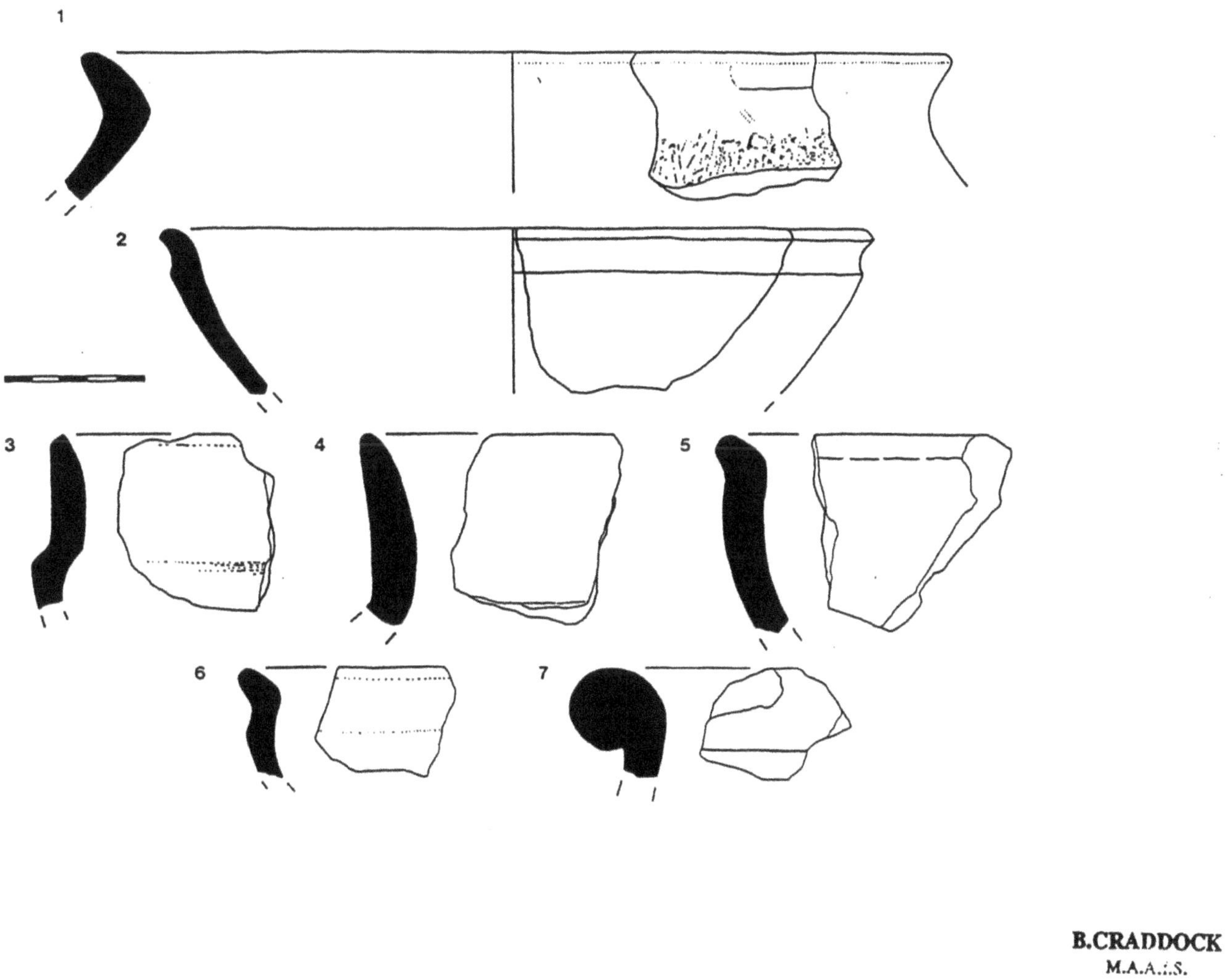

Drawing No 4 survey pottery, San Cristobal, scale units 1cm

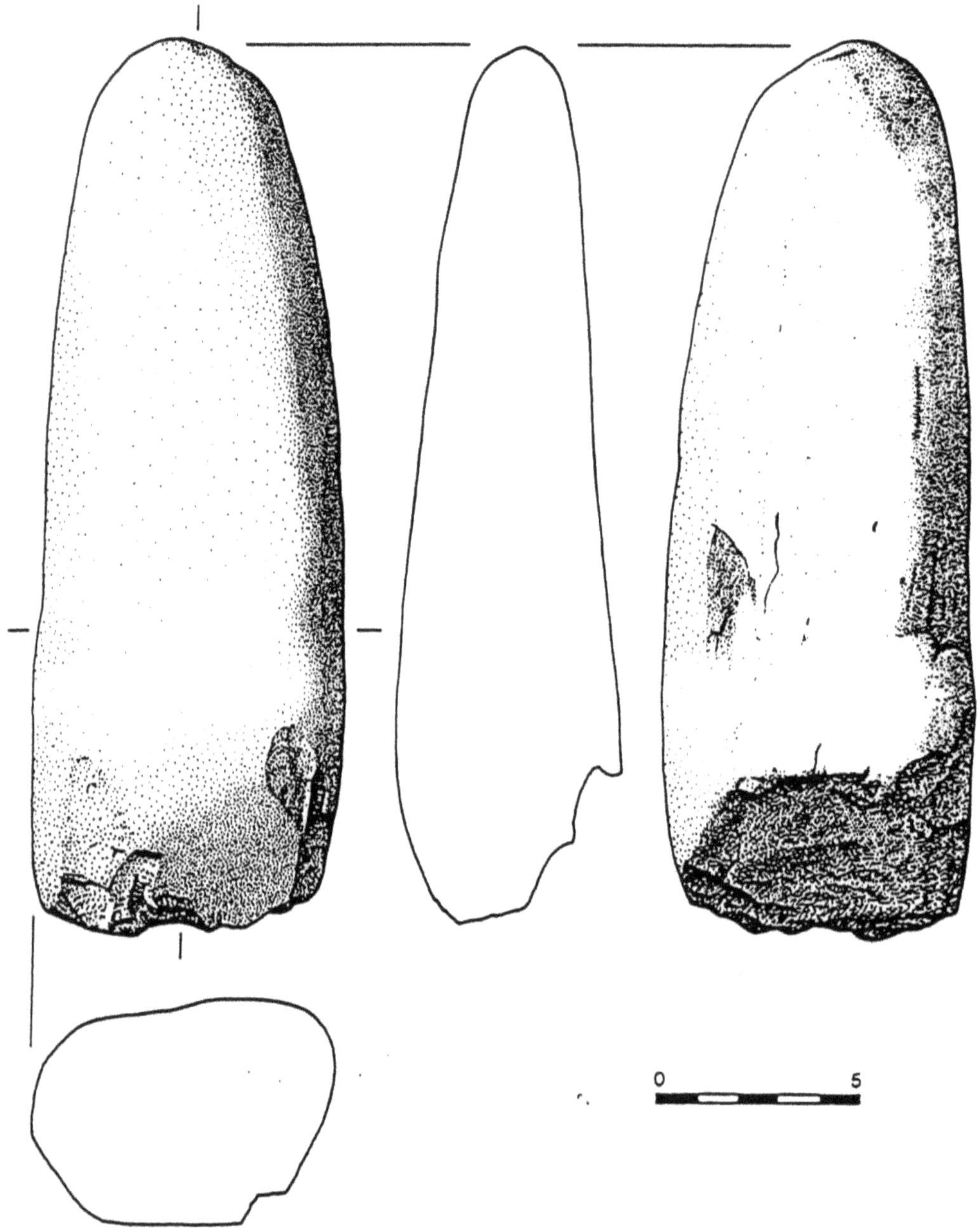

Drawing No 5 stone hammer, San Cristobal, scale units 1cm

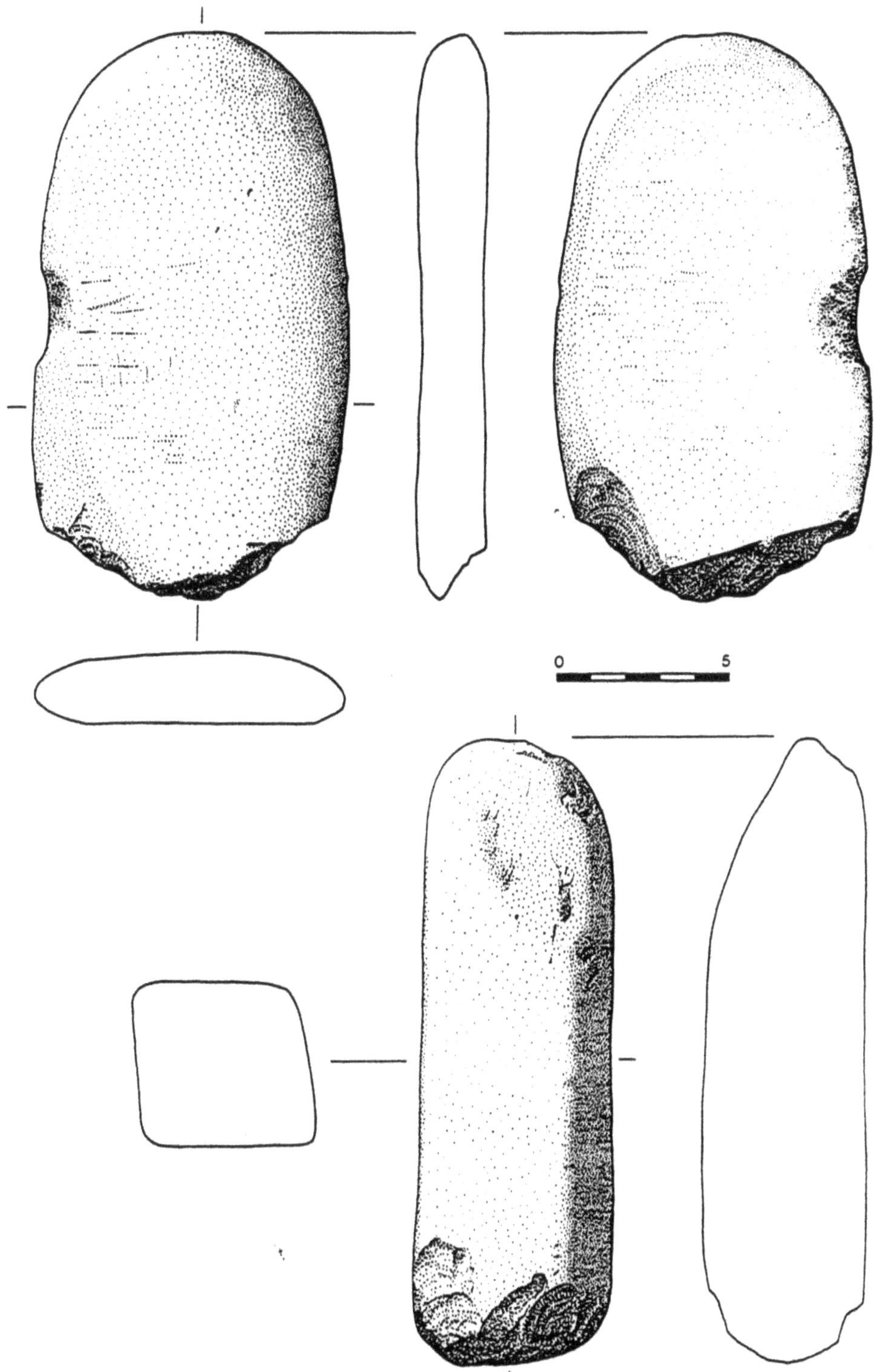

Drawing No 6 stone hammers, San Cristobal, scale units 1cm

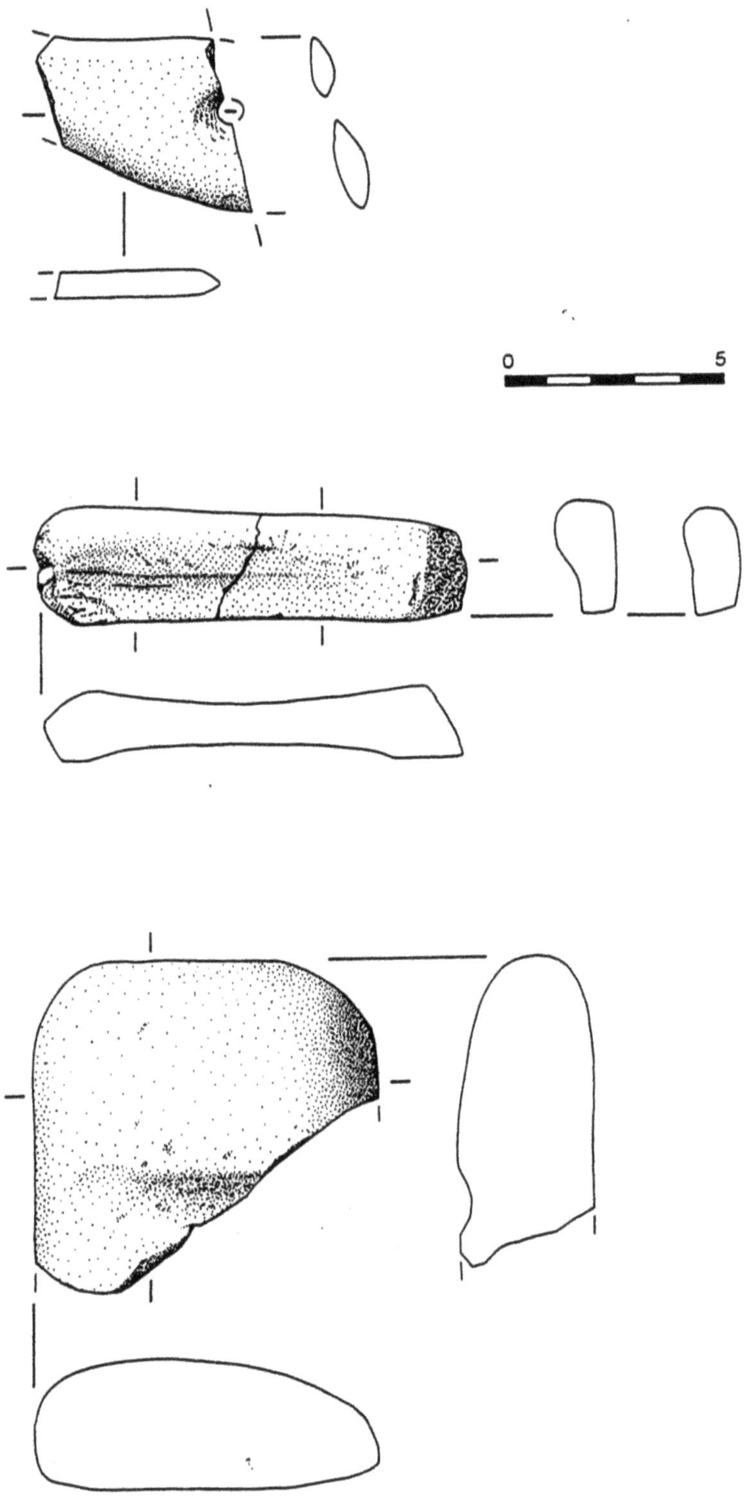

Drawing No 7 worked stones, San Cristobal, scale units 1cm

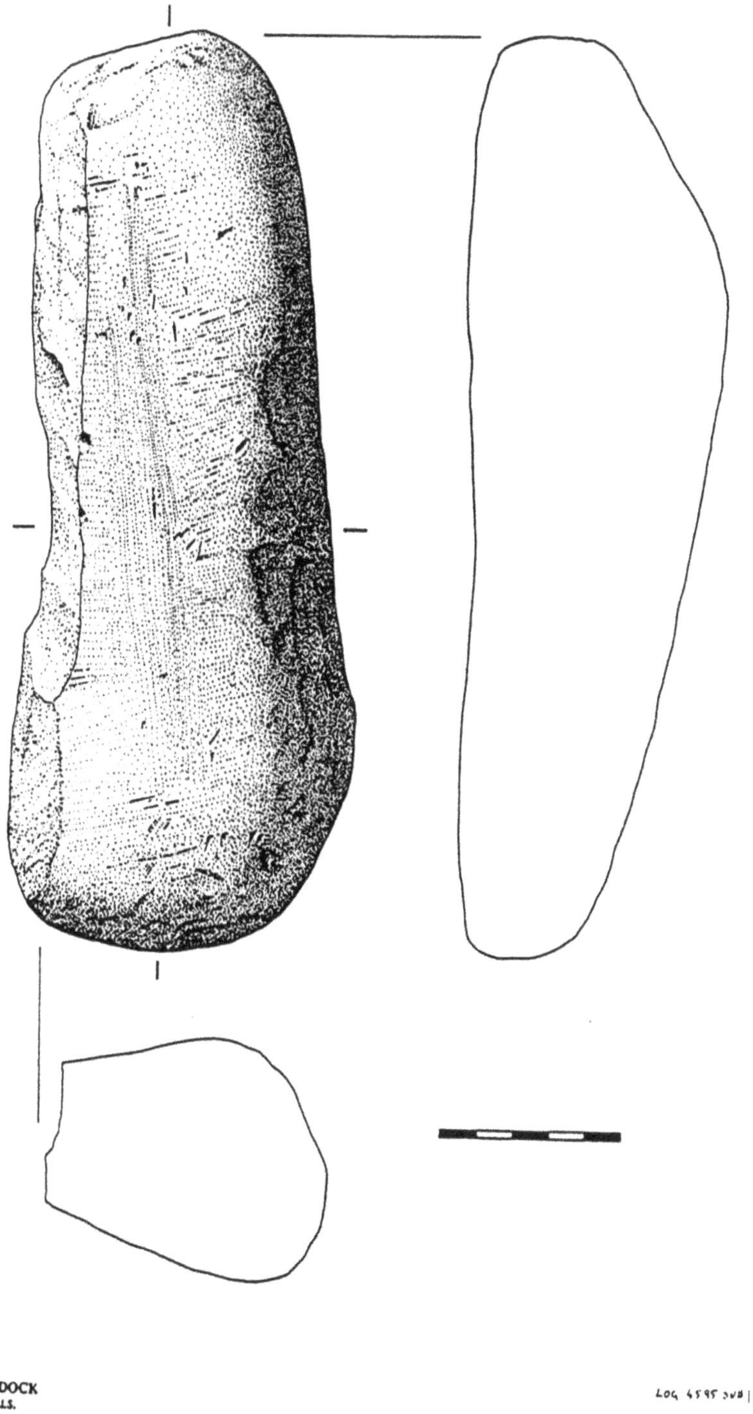

Drawing No 8 stone hammer, San Cristobal, scale units 1cm

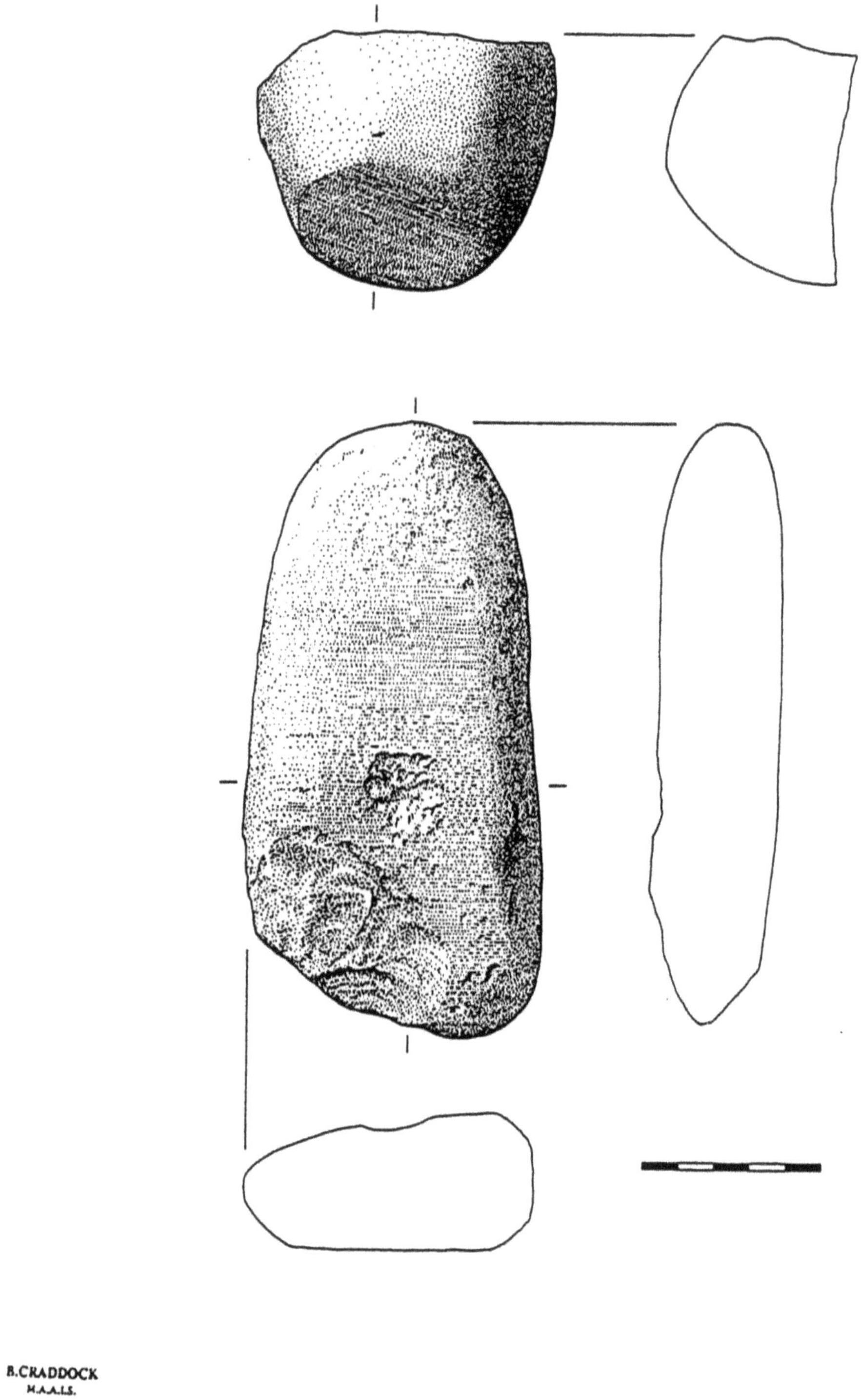

Drawing No 9 stone hammers, San Cristobal, scale units 1cm

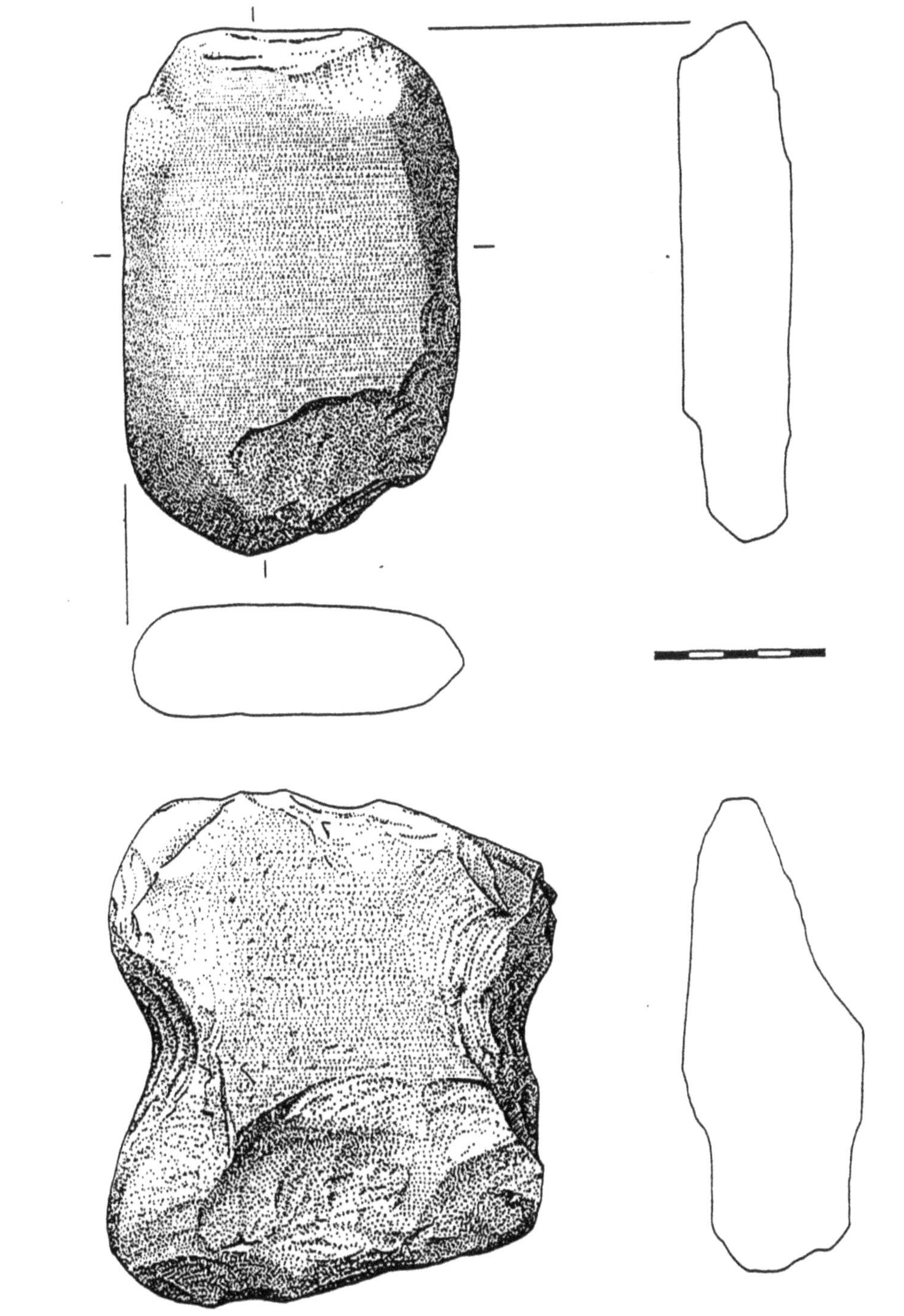

Drawing No 10 stone hammers, San Cristobal, scale units 1cm

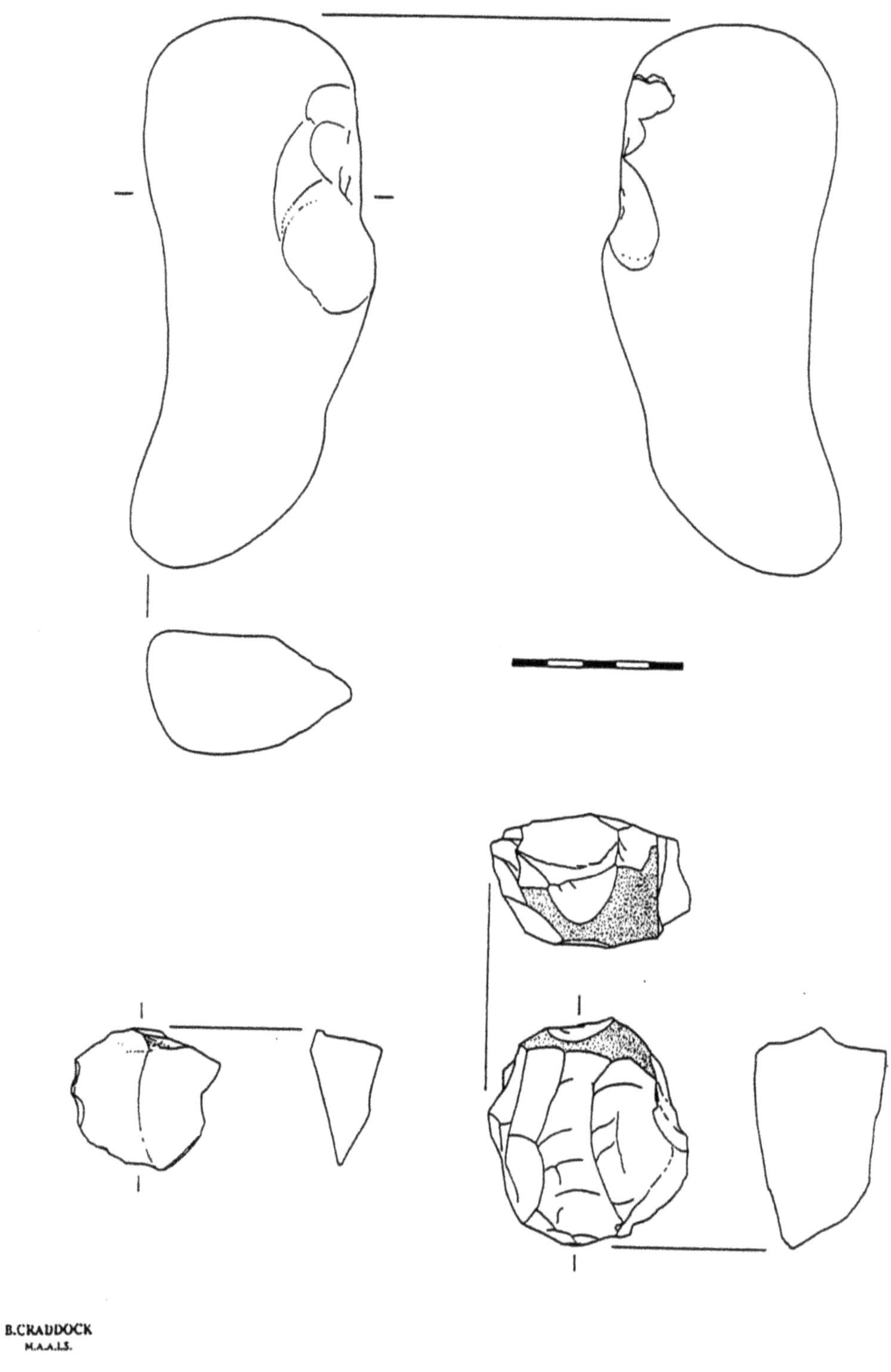

Drawing No 11 stone and flint, San Cristobal, scale units 1cm

VISIT NO 5 **LA MINA PLASENZUELA**

PROVINCE: CACERES

CO-ORDINATES: WEST: 06° 02' 00"- NORTH: 39° 23' 20"
Provinical Map CACERES 1:200.000 1989 2a ed.

Photo 6 La Libre Mine Area - Plasenzuela - looking south west

The Plasenzuela granite is a stock with dykes of pegmatite and aplite granite present (Molero, 1993, 62). The stock is an oblong shape approximately five kilometres wide by eight kilometres long. The main problem encountered at this site is that of the eight listed mines near Plasenzuela, none are listed as cassiterite-wolframite mines (Ochoa,1987, 100-1). The following ores are listed as being present: quartz, galena, sphalerite, pyrite, chalcopyrite, and jamesonite ($Pb_4FeSb_6S_{14}$) in the eight mines. The main ores in the Plasenzuela mineral area are galena and sphalerite. Cassiterite and wolframite do not seem to be listed as present

Both Domergue and Edmondson state that Plasenzuela was a major Roman mining district for the production of lead, silver and copper. They also agree that the Roman mine workings (dated to first half of first century BC) are quite numerous and cover a distance of six kilometres along the Western side of the Plasenzuela stock to the North West of the village of Plasenzuela (Edmondson, 1987, 226).

The only Plasenzuela mine area which I actually surveyed during my visit was probably the mine area called La Libre (Photo 6) lying about one and one half kilometres to the North East of the village of Plasenzuela near the eastern edge of the granite stock. There was no sign of any of the modern mine workings at La Libre being currently in production.

The modern mine workings seem to have been a shaft and gallery mining system. It would be difficult to give a precise date to any of the shafts and galleries as I did not enter them for an inspection. The underground workings do not seem to have been very extensive nor very deep, if one is to go by the small size of the surface waste heaps next to the shafts investigated.

It is possible that the underground gallery waste material was not being brought to the surface but being dumped into the old worked out underground galleries as new galleries were being opened up.

I surveyed a small part of the Plasenzuela area near the La Liebre Mine area for about half a day and found no surface evidence of either ancient mining or smelting activities. On the basis of my own findings I do not think that I would return to this area for further survey work. The minerals cassiterite-wolframite do not seem to have been present in this area. The lead and silver deposits in the Plasenzuela mine area might necessitate another look in the future.

The dilemma with the Plasenzuela visit is that the descriptions of the remains of Roman mines and mining artefacts by both Domergue and Edmondson do not agree with my visit. I suspect that I did not locate precisely the area described by both Domergue and Edmondson.

Since doing my survey at Plasenzuela I have found an article covering archaeological survey and excavation work of the Chalcolithic period within the Plasenzuela granite stock area (Navascues, 1991, 11-26). The article on the survey and excavations work has made me realize that the Plaenzuela area has a long and rich archaeological history. Another realization is that I had visited only a very small area of the Plasenzuela granite stock region. Therefore a future and more extensive survey visit to other areas and mines of the Plasenzuela granite stock region is a necessity.

VISIT NO 6 **LA MINA DE TRES ARROYOS**

PROVINCE: BADAJOZ - MUNICIPALITY ALBURQUERQUE

CO-ORDINATES: WEST: 07° 07' 30" - NORTH: 39° 15' 00"
Hoja 727 ALBURQUERQUE 1:50.000 1963 2a ed.

Photo 7 View of Tres Minas Mine/Smelting Area
looking to the North West (Slag location area in the middle near foreground) by rocks

La Mina de Tres Arroyos sits on the south eastern contact side of a granite batholith which runs N/W to S/E from the Rio Zapaton above the Embalse de la Pena del Aguila to the region of Castelo de Vide in Portugal. The contact rock with the granite batholith is slate (Barrena and Isidro, 1987, 78).

La Mina de Tres Arroyos is a Cassiterite/Wolframite mine covering a total of 750 hectares, located approximately 10 kilometres to the North West of the town of Alburquerque in Badajoz Province. The actual Tres Arroyo mine area is located between the Arroyo de Fraile and the Arroyo de Alconero. In the modern period the Tres Arroyos Mine was worked from 1959 to 1960, and then from 1962 to 1969. The mine produced yearly 5 tons of concentrate, containing 65% tin ore oxide and the rest tantalum and niobium. The main effort of the modern mine took place in three open cast ore quarries, each approximatly 130 meters by 60 meters and about 12 meters deep (Molero, 1993, 99).

The Mina De Tres Arroyos area is mentioned as having ancient Roman mining and smelting workings. Roman pottery of the Terra Sigillata type has been found at the Tres Arroyos site and it is considered that the cassiterite was worked out of the quartz veins present in the local granite and then possibly smelted, from the end of the 1st century BC to the start of the 2nd Century AD (Domergue, 1987, 16-17).

A day and a half was spent surveying the Tres Arroyos Mine site, which is a short time considering the total land area covered by the mine. It was extremely hot August weather at the time of the survey (with temperatures as high as 44 °C). The modern mine open quarry areas were found as was the area of the modern crushing plant and former mine living areas. None of these areas provided any evidence of ancient mining or smelting.

Photo 8 Tres Minas Smelting Area
slag location area by 20 cm measuring rod

The area directly to the South-West and West of the modern crushing area was extensively covered with numerous small pits and crushed quartz heaps next to the pits; very extensive low shrub vegetation hindered any real in depth survey of this area, (unless one was to crawl around on one's hand and knees). A lot of the small pits had been eroded or almost weathered back to the original ground surface level.

A cleared ploughed field area to the South-East of the abandoned modern mine crushing area might represent the ancient "smelting furnace remains" mentioned by Domergiue (Domerque, 1987, 16-17).

Five pieces of porous slag where found, three pieces were collected for laboratory analysis in London and two pieces were left in sital on the surface (Photo 8). Twelve pieces of pottery were collected from the surface area of the proposed smelters, but most pottery was also left in sital. Most of the ploughed field area had some pottery on it,and most of the pottery seemed to be a mix of modern and possiblly Roman. No evidence of any stone mining or crushing tools were found.

For all the survey time spent at the Tres Arroyos site only the above area yielded slag and pottery on the surface. The Tres Arroyos site still remains very interesting and more time should be spent there on further surface survey work of the general area.

A small excavation should be undertaken in the probable smelting area where the five pieces of surface slag were found.

The alleged smelter area's seem to have been dug into the South West slope of the small hill of the ploughed field. One proposed smelting area is extremely disturbed and has possibly been excavated in the past, but in what seems to have almost been a treasure hunting manner.

Twelve pieces of pottery from this site were examined by Alonso Rodrigvez Diaz and Ignacio Pavon Soldevia from the Universidad de Extremadura in Caceres and were considered to be a mixture of Roman and Modern.

The three pieces of slag found at this site were cut, mounted and analysed by Scanning Electron Microscope at the Institute of Archaeology University College, London. The analytical results showed that the slag was probably Iron slag, as no traces of tin, tantalum or niobium were found within the slag matrix of the pieces sampled (Samples No's 13, 14, 15, page 136).

VISIT NO 7 **MINA DE SANTA MARIA**

MUNICIPALITY: PEDROSO DE ACIM PROVINCE: CACERES

CO-ORDINATES: WEST: 02° 42' 40" - NORTH: 39° 50' 45"
Hoja TORREJONCILLO 1:50.000 1a edicion 1946

Photo 9 General View of Mina de Santa Maria looking to North West

The Santa Maria Mine is a irregular ancient alluvial deposit, possibly laid down in the late Tertiary Era, which has the cassiterite ore grains and accompanying quartz gangue lying in clay and found at a depth of 20 meters. Some of the cassiterite grains found normally encased in quartz gangue at the Santa Maria Mine have been up to three centimetres in size (Barrena and Isidr, 1987, 51-2).

Cassiterite was discovered at this site in 1957. From 1957 to 1967 the Santa Maria mine area was mined by local individuals but only worked seasonally due to water shortages. The water was needed for the extraction of the cassiterite from the clay and crushed quartz gangue (Penhallurick, 1986, 97). From 1967 to 1980 the Santa Maria mine was worked by the mining company la Sociedad Anonima Estanifera Extremena. During this period of time the total production of the mine was 400 tons loaded with a concentration of 70% tin opxide ore. Mining ceased at the Santa Maria mine in 1980 (Molero, 1993, 87).

I surveyed the Santa Mira mine area on two occasions spending in all about one day on the site. I also surveyed additional surrounding areas to the south and east of the modern mine works. During the survey period I encountered no evidence of ancient mining or smelting remains and found no stone mining tools nor fragments of any ancient pottery. The modern open cast mining methods had been very destructive at this site.

It is possible that there might have been a small centre of ancient mining within the modern open cast mine area or near the dry river bed located approximately one hundred meters to the east. The dry river bed runs in a north east to south west direction and has exposed quartz veins in its banks but no visible cassiterite ore within these veins.

The depth of the cassiterite deposits at 20 meters and the lack of any surface evidence of ancient mining and smelting argue very much against the Santa Maria Mine area being in use in antiquity. Therefore I do not think that it will be worthwhile returning to the Santa Maria Mine for any further survey work.

VISIT NO 8 **MINA CALERO**

MUNICIPALITY: PERALES DEL PUERTO

CO-ORDINATES: WEST: 06° 40' 00" - NORTH: 40° 08' 15"
Provincial Map CACERES 1:200.000 1989 2a ed.

Photo 10 General View of Mina Calero Mine Area looking to North West

This area was visited as it was listed in The Mines of Lusitania as an area of alleged Roman tin mines (Edmondson, 1987, 225). The Perales Del Puerta area is not covered in any detail in the latest book on the Mines of Extremadura, nor does this book contain any information on any old or prehistoric mining remains within the Perales del Puerta area (Molero, 1993, 61-121).

Most of the mines found within the Perales Del Puerta area seem to be Wolframite mines with only one or two of the mines having cassiterite accompanying the wolframite (Ochoa, 1987, 93-4). All the mines are located on the edges of the Hoyos Batholith which runs in a SE to NW direction and carries on into the Portuguese province's of Beira Baixa and Beira Alta.

The area I surveyed was on the North/East edge of the batholith from below Perales del Puerta to below the town of Aceba. It is on this edge of the batholith that most of the modern mines for Wolframite are located.

The South/West edge of the batholith is the area of the tin mines, from just below the town of Valverde del Fresno to just below the town of Cilleros. I did not look at this area during my survey work.

It seems now after looking at all the available geological books and maps, that I spent most of my time examining only the Wolframite mining areas. Making the mistake of surveying a wolframite ore area most probably explains the complete lack of any ancient mining or smelting remains on the north/eastern edge of the Hoyos batholith.

If I can locate more precise detailed geological information of cassiterite mineralization in the Hoyos area along the north/eastern edge of the batholith I will return to the Hoyos area for more survey work.

A future survey visit to the now known cassiterite area on the south/west edge of the Hoyos batholith should offer a better chance of possibly finding ancient worked cassiterite mines.

VISIT NO 9 **MINA LA LAPA**

MUNICIPALITY: HERNAN PEREZ

CO-ORDINATES: WEST: 06º 29' 20"- NORTH: 40º 12' 20"
Provinical Map CACERES 1:200.000 1989 2a ed

Photo 11 General View of Mina La Lapa - looking to North West small open cut trench mine in foreground

This mine was listed as being a cassiterite/wolframite mine area (Guzman,1993, 47). The La Lapa Mine is located on the South/Eastern edge of a batholith, granite area which runs in a North/West to South/East direction. The actual mine mineralised area is situated on both the East and West side of the Rio Arrago approximately four hundred meters straight north of the bridge where the number 513 road crosses over the Rio Arrago between the pueblo's of Hernan Perez and San Juan.

On my first visit to the mine area I had a great deal of trouble locating the exact mine site. It was only due to help from some local people in the pueblo of Hernan Perez that I was able to find the mine area.

The mine area had last been worked in 1976. There was not very much remaining in the way of mine buildings or actual modern mining evidence.

The various mineralised veins of the La Lapa Mine run in a direct north south direction (Molero, 1993, 70-3).

All that I found in surveying this mine site was the remains of modern trench exploration mining mostly machine dug and only to a depth of one to two meters.

The very dense wild vegetation in some parts of the La Lapa mine area made a through foot survey very difficult, if not impossible. The possible danger of hidden or covered mine shafts within the thick vegetation did not encourage me to try to explore those areas in any great detail.

The area that I surveyed was the mineralised mine area along the northern side of the Rio Arrago. The survey produced no evidence of ancient mining or smelting. I did collect some mineral samples, which when later analysed proved to be Wolframite.

In 1993 I found out from a more recent and detailed published report that the La Lapa mine was only a wolframite mine and that cassiterite had not been mined or found at this site (Molero, 1993, 73).

VISIT NO 10 **LA MINA DE DESQUITES**

MUNICIPALITY: TORRECILLA DE LOS ANGLES

CO-ODRINATES: WEST: 06° 26' 00" - NORTH: 40° 15' 25"
Provincial Map CACERES 1:200.000 1989 2a ed.

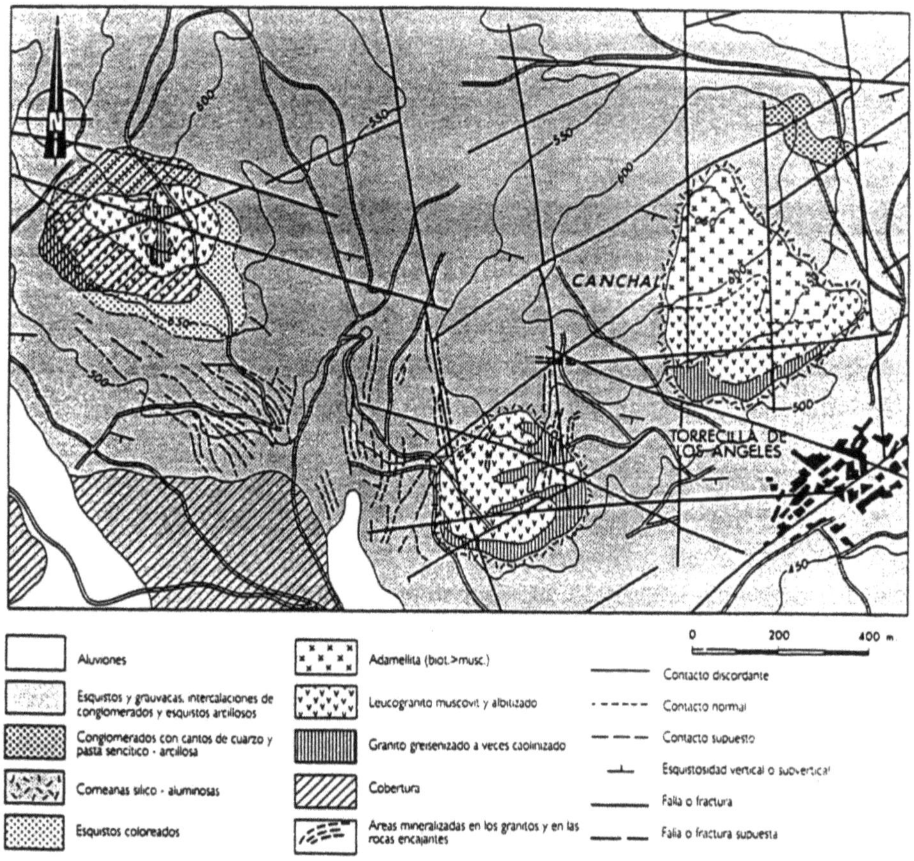

Fig 8.3 Geological Mine Plan of the Apophysis Granite De Torrecilla De Los Angles (North Caceres) Molero, C.A., 1993, 74

The mine of Desquites just to the north east of the village of Torrecilla de Los Angles is listed as being a tin mine in modern mine publications (Guzman, 1983, 47).

The Desquites mine area is formed geologically by three igneous off shoots or apophyses of a supposed larger deeper igneous body. The three apophyses constitute the Desquites Mine area. The granite is a light coloured granite (leucogranite) with mica and formed by albitisation and muscovitization (Molero, 1993, 73).

The minerals appearing at the Desquites Mine are cassiterite, wolframite and arsenopyrite.

The Cassiterite appears in small stockwork veins in the granite, kaolinised granite and surrounding gravels and conglomerate rock.

The kaolinised granite areas are very soft and could easily have been worked without either explosives or ore crushing mills (Molero, 1993, 73).

A day and a half were spent at the Desquites Mine which does not cover a very large area. The largest of the three apophyses granite areas which constitute the Desquites Mine being only 500 metres by 300 metres and having a kaolinised mineral zone of only 300 metres by 20 metres (Molero, 1993, 73).

Photo 12 General View of Mina De Desquites looking to the South, showing the disturbed nature of the mine area

With all the above points taken into consideration one would expect to find some evidence at this mine of ancient mining, ore preparation and even possibly smelting activities. Unfortunately the work of the modern mining activity this century seems to have destroyed or obliterated any ancient remains if they did exist (Photo 12).

The Desquites Mine represents a tin mineral area that fulfils all the possible criteria for being an ancient mine area. It is not a large mineral area and all the mineral occured near the surface and the kaolinised granite areas within which some of the cassiterite appeared was easy to mine.

This site produced a few sherds of pottery one or two of which may be of Bronze Age origin. Also found on the surface were two supposed fragments of stone querns which are impossible to date. Also found was a fragment of a possible stone hammer.

More time should be spent surveying at this mine. Though the Desquites Mine did not produce good results on my first visit, the non-alluvial nature of the area and the easy to work mineralization of the cassiterite point to a tin mineralized area that could have been worked in antiquity.

VISIT NO 11 **MINA COMOHAL HUBARBA Y HORRERA**

MUNICIPALITY: POZUELO DE ZARCON

CO-ORDINATES: WEST: 6° 24' 00" - NORTH: 40° 09' 20"
Provincial Map CACERES 1:200.000 1989 2a ed

Photo 13 General View Mina Comohal Hubarba y Horrera looking to the North East in foreground small machine exploration trenches

The mine was located about five kilometres to the North-East of the Pueblo of Pozuelo De Zarcon on the road North to the pueblo of Santa Cruz de Paniagua. The mine was located to the east side of the road.

The modern mine area lies in the slate area adjoining the north/western edge of a very large porphyry granite belt/field of about two hundred kilometers in length (approximately East-West) and about eighty to ninety kilometers in width (approximately North-South). This large granite belt runs from the region of Pozuelo de Zarzon in the province of Caceres to the north of Madrid in the province of New Castile and represents most of the mountains of the Central Cordillera.

This mine did not appear in any of the references that I had on modern tin mine locations in Extremadura. I was told of the location by a man from the pueblo of Pozuelo De Zarcon. I am still not sure if it was an actual mine site or just an area of quartz with a lot of tourmaline present. There is a mine in this area listed in the Mapa Geologico De Extremadura (Ochoa, 1987, 92). On this map and the accompanying guide the mine location is put in the Termino Municipal, Santa Cruz de Paniagua. This mine is listed as a vein mine containing quartz, cassiterite, arsenopyrite and micas, but with the vein direction unknown. It is very possible that this is the mine that I visited.

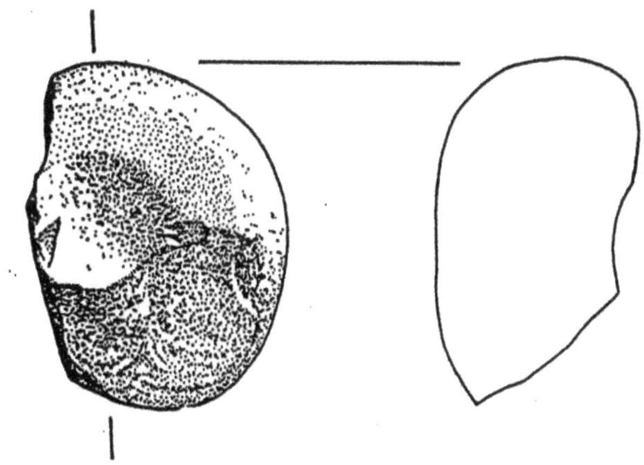

Drawing No 12 grinding / hammer stone fragment scale 1/2-1
Mina Comohal Hubarba y Horrera

There were a lot of shallow trenches about half a meter deep which looked like they had been recently made by a narrow shovel of a mechanical digger in the quartz field area. I spent a day walking around the area but found only one definite archaeological find: an almost complete fragment of a finely shaped small round crushing stone, about five centimetres in diameter (Drawing No 12). It appearedto have been used for crushing/grinding on one centre face and around its outer diameter. It is not clear if this stone tool was used for ore preparation or other activities.

Due to the large area of this site another visit and more survey time is needed, but I am not sure that more time spent at this supposed mine site will produce any postive evidence of ancient mining or smelting.

VISIT NO 12 **LA MINA CARRASCAL**

MUNICIPALITY: VILLA DEL CAMPO

CO-ORDINATES: WEST: 6° 26' 00" - NORTH: 40° 08' 00"
Provincial Map CACERES 1:200.000 1989 2a ed

Photo 14 Gallery Entrance La Mina Carrascal - looking to East

Again this mine location and name was told to me by a man in the Village of Pozuelo De Zarcon. Listed in this Mine area in the Mapa Geologico De Extremadura are two tin mines, the one that I visited is most probably the one listed as No 48. This vein mine lies in "fleckshiefer" spotted slate and consists of quartz, cassiterite and tourmaline lying together in veins trending in a North/East-South/West direction (Ochoa, 1987, 95).

It took a long time to find this mine as it is not a very large affair and most people that I asked about its location did not seem to know anything about it. This suggests that it was worked some time ago, perhaps no more recently than the early part of this century.

When I finally located the mine I could only note its position and state that it is a gallery mine system probably of not too great a length or depth. There had been a substantial roof rock fall at the entrance of the gallery so I decided that it would not be too wise to venture in (Photo 14).

The mine gallery height and profile seemed to represent a late 19th or early 20th century mine working and I found no pottery around the mine entrance or in the adjoining waste heap area. The waste heaps were not large so I do not think that the mine gallery is very extensive. This was not a very interesting site for early archaeometallurgy survey work. I will probably not return to this mine for any future work.

VISIT NO 13 **MONTEHERMOSO**

MUNICIPALITY: MONTEHERMOSO

CO-ORDINATES: WEST: 6° 20' 10" - NORTH: 40° 06' 00"
Provincial Map CACERES 1:20.000 1989 2a ed

Photo 15 Entrance Shaft/Gallery Montehermoso Mine - looking to North East

This mine is located to the South West of Montehermoso off the route C-204 to Plasenca. This Montehermoso mine is mentioned in a couple of publications as being a tin mine (Gumel, 1981, 9).

I again had a lot of trouble locating the Montehermoso mine as most people I asked did not seem seem to know of its existence or location and did not know which minerals had been mined in the area.

The mine is a vein mine running in a NE-SW to E-W direction and is located in spotted slate and granite. The composition of the vein is quartz, wolframite, cassiterite, pyrite, arsenopyrite and tourmaline (Ochoa, 1987, 94).

When I did locate the mine it was a shaft/gallery mine system and quite deep (20-25 meters).

The two shafts that I saw were quite large at least (5 meters in diameter) and both filled with water (Photo 15). So it was impossible to see the shapes and sizes of the galleries leading off from the shafts. The waste heaps by the mine shafts were not very extensive. I was told that the mine had been worked for Wolframite and Tin for about 40-55 years (1920's-1960's); to my mind the visible external waste heaps did not seem to represent such a long period of mining. It is possible that some of the mine waste heaps have been used for road building activity since the mine closure.

I searched the waste heaps and the surrounding area but found no evidence of early pottery or crushing tool remains. I did not spend much time at this site, as I felt that the mine was too deep to have been a ancient mine, and that any more time spent at this site would be wasted time.

VISIT NO 14 **MINA TELBA**

MUNICIPALITY: CASAS DE MILLAN

CO-ORDINATES: WEST: 6° 15' 00" - NORTH: 49° 47' 00"
Provincial Map CACERES 1:200.000 1989 2a ed

Photo 16 General view Mina Telba area, looking East

This mine is located south of the road between Casas de Millan and Serradilla about 10 kilometres to the East of Casas de Millan. It is not difficult to find as it was a medium sized open cast mining operation 20-30 meters in depth and approximately 100 meters in diameter. The mine closed down in the mid-80's due to falling prices of tin after working for about 20 years. The owner said the mine site at Telba will be opened up again if the price of tin ever reaches the levels achieved in the early to mid-80's.

The geology of this mine is very different from most of the tin mines found in the Province of Extremadura. There is no granite visible or known within the mine area. The closest granite is a small outcrop of granite (3 kilometres long by 11 kilometres wide) located approximately 12 kilometres to the North West of Telba by the village of Pedroso de Acim (Fig 8.4).

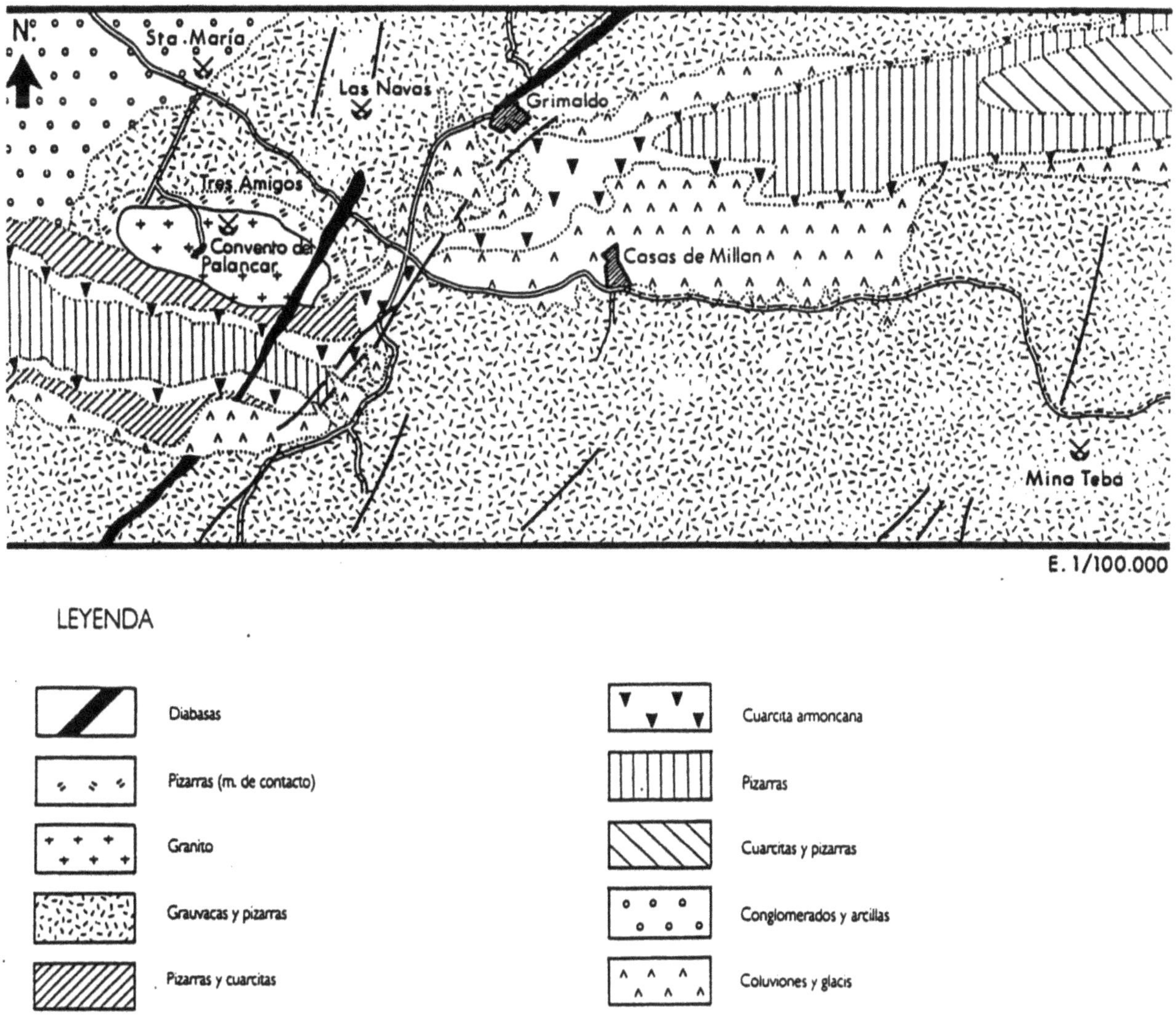

Fig 8.4 Mina Telba - Santa Maria - Tres Amigos
Molero, C.A., 1993, 86

The mineralisation at Telba consists of quartz, cassiterite, tourmaline and micas in irregular stockwork veins with the three principal vein directions being N-S, N25-E, and E-W (Ochoa, 1987, 98). The host rocks is arenaceous (graywacke) rocks (basically a hard drark grey coarse graimed sedimentary sandstone) (Bates and Jackson, 1987, 289) slate rocks also form part of the host rock complex.

I surveyed around the open cast mine, spending almost a complete day at the site. To the South West of the open cast area I found what could possibly be the fragment of a stone hammer (Drawing No 13) and some very fragmented pottery shreds which might derive from the Bronze Age or Roman period.

I showed the pottery to Alonso Rodriguez Diaz at Caceres University and he said that the pottery shreds were too fragmented and unidentifiable to use for absolute postive dating.

It would be difficult to associate the pottery with any mining activities as the stone hammer and pottery fragments could represent a small pre-historic hunting site or settlement site. The mine area of Telba will require another visit and more surface survey effort.

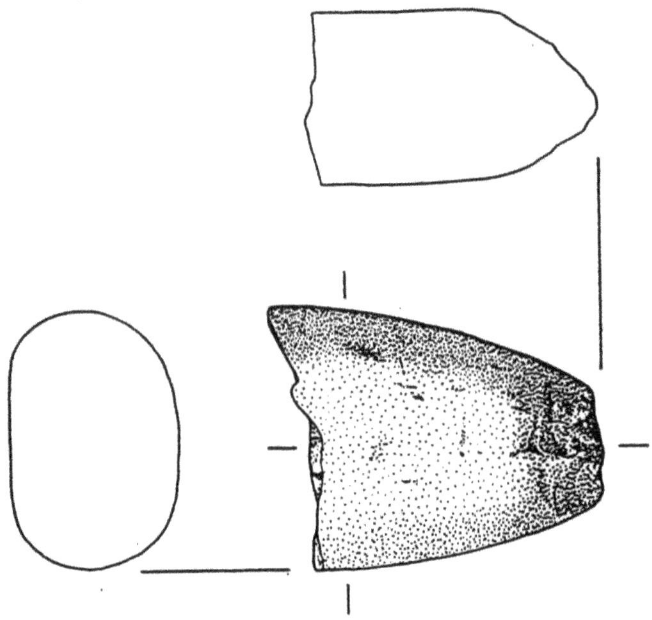

Drawing No 13 fragment stone hammer, Mina Telba scale 1/2-1

I also collected a small mineral sample from a surface quartz vein in the South West upper area of the open cast which I thought might be cassiterite. I analysed this sample on the SEM and it proved to be cassiterite with 65%-75% Sn present (Sample No 4, page 137).

The cassiterite mineralisation at this site would have been available and usable in the Bronze Age due to its surface location in visible quartz veins.

Visit No 15 **MINA CLAUDINA**

MUNICIPALITY: MORILLE

CO-ORDINATES: WEST: 5° 42' 00" - NORTH: 40° 48' 05"
Provincial Map SALAMANCA 1:200.000 1984 2nd ed

Photo 17 General view of Mina Claudina area - looking to East

Located in this area to the North and West of Morille are three mines the San Andres, the Mundaca and the Claudina containing the minerals cassiterite and wolframite. The mines were open cast pit and vein workings (Junta De Castilla Y Leon, 1988, 103).

I visited the Mina Claudina which produced the minerals Scheelite and cassiterite. The Mina Claudina stoped working in 1983; it had started working in 1942-1943. I spent most of a day at this site doing a fairly thorough surface walking survey. No evidence of early mining or mineral preparation or smelting was observed. There were large amounts of good surface tourmaline minerals in the mine areas that I looked at.

After leaving the Mina Claudina I tried to find the mines of Pili near San Pedro de Rozados and El Cubito near El Vecino. The San Pedro de Rozados mine area had been mentioned as being worked in Roman times (Edmondson, 1987, 228).

The El Cubito mine is described as a large modern opencast working for tin which is no longer in production (Junta De Castilla y Leon, 1988, 116-20).

I had no luck in locating either mine, only a couple of shafts of about 20-30 meters depth which seemed to have no side galleries off them and could have been disused mine shafts or more likely old disused well shafts.

Most of the major roads go out from Salamanca in the form of spokes from a central hub of a wheel. The land between these hubs consists of extremely large farming estates. These large ranching/farming estates are fenced off and impossible to enter and it is on the ranch land that the cassiterite mine areas were located which I was trying to visit.

A further problem which made the search difficult was that some of the estates where I felt the mines might be located were farms for raising fighting bulls for the ring. Faced with these obstacles I finally had to give up the this part of the search.

VISIT NO 16 **MINA GOLPEJAS**

MUNICIPALITY: GOLPEJAS

CO-ORDINATES: WEST: 5° 54' 42" - NORTH: 41° 00' 03"
Provincial Map SALAMANCA 1:200.000 1984 2nd ed

Photo 18 View of Mina Golpejas mine crushing-washing area with waste heaps - looking to North

Golpejas is another mine that is mentioned as having been worked in the Bronze Age and possibly Roman times (Domerque, 1990, 129-30). However no firm archaeometallurgical surface evidence of this was found during my survey work at Golpejas.

Golpejas was a very easy mine to find as it was a very extensive modern open cast mining operation which has left extremely large white waste heaps (Photo 18). The Golpejas mine stopped working in 1985. The mine had been worked for about 30 years. It was one of the major tin producing mines in Spain during its working life.

The production of the modern mine was large enough to keep thirty six shaker tables working after the granite had been crushed by three large rotary ball mill crushers. The mine was a large open cast operation consisting of two pits called the "Tita" and "Bellita" which produced tin and tantalum (Junta De Castilla Y Leon, 1988, 112-14).

In Golpejas the Cassiterite ore appears in the form of very fine grains accompanied by the mineral Columbo-Tantalum (which contains Niobium and Tantalum) and small amounts of sulphide.

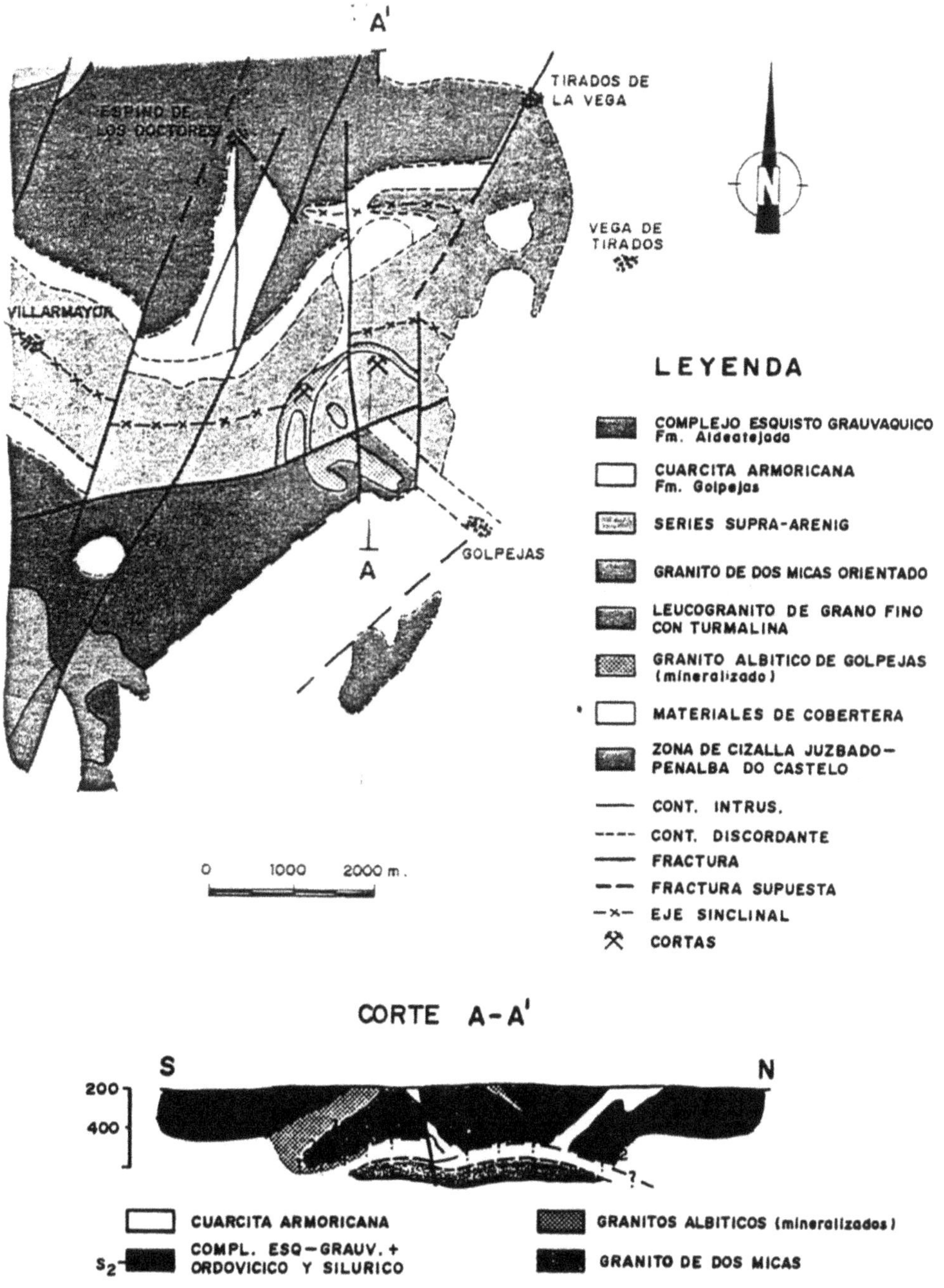

Fig 8.5 Geological Plan of the Golpejas Sn ore deposit
(Iglesias M. y Conzalo F., 1981) Junta de Castilla y Leon, 1988, 113

These are scattered through a sill of light coloured fine grained granite which is heavily folded. The granite sill is approximately 2400 meters long, has a semi-elliptical shape and is approximately 30 metres deep (Fig 8.5). The granite sill is emplaced in a metamorphosed slate formed during the Ordovician-Silurian Periods 500 - 395 million years ago (Junta de Castilla y Leon, 1988, 112-14).

It is thought that the granite has been altered by hydrothermal kaolinization and greisenation (a process in which feldspar and muscovite are converted to an aggregate of quartz, topaz, tourmaline and lepidolite by the action of water vapour containing fluorine) (Bates and Jackson, 1990, 290). The richest tin areas were located in the top area of the granite which had been kaolinized (containing approximately 2.5 kg of Sn per ton of material mined).

Photo 19 Rock which contained Cu-Sn Minerals, Mina Golpejas, 20 cm measure rod

The Golpejas ore deposit constitutes a model of tin and tantalum desposition which is almost unique and appears in only one other place in the world: in the mineral deposit of Dajishan in China (Consejeria de Economia y Hacienda/Castilla y Leon, 1988, 112).

Since 1992 I have spent a total of 5 days surface surveying at the Golpejas Mine trying to find evidence of early surface archaeometallurgical remains but not with much success.

During my survey there were three locations that I found of interest within the Golpejas mine area. I gave these locations letters Point A, Point B, and Point C.

At Point A, I found a mineral which contained in various percentages copper and tin. This mineral appeared in quartz boulders which had been cleared and dumped on waste heaps as part of the clearance of the overburden of the open cast pits. In the analysis section (Pages 138-142) I will present the geology and analysis of this mineral (six separate samples) with which one may have been able to make "natural bronze" directly from smelting.

At Point B called in Spanish Tierras Viejas (The Old Earth), I found quite a bit of surface pottery most of it probably 19th or 20th century but also with the possibility that a few sherds might be Roman and Late Bronze Age. When I first surveyed Point B it was just after the summer crop of wheat had been harvested. I was therefore able to walk the fields of this area quite freely.

During two return visits to Golpejas I was not able to survey the Tierras Viejas again due to fact that both visits were made during the planting and growing seasons. The Tierras Viejas area should be surveyed again as it seems to hold the best promise of yielding further archaeological surface pottery sherds.

At Point C, about 250 meters South of Road C 517, direction Golpejas-Vitigudino kilometre 24.5, there was again an area that provided quite a bit of surface pottery. Most of this pottery was classified as 19th and 20th century (Dr Angel Esparza Arroyo - Universidad de Salamanca - Departamento de Prehistoria).

The Golpejas area has not to date provided any firm evidence of early mining or ore preparation. The cassiterite ore that had been mined in the modern open cast pit operations was most probably too deep to have been worked during the Late Bronze Age or the Roman period.

However the copper/tin mineral which I found in the waste heaps at Point A could have been used during the Late Bronze Age or Roman period, but no evidence to support this theory has been found to date during my survey work in the Golpejas area.

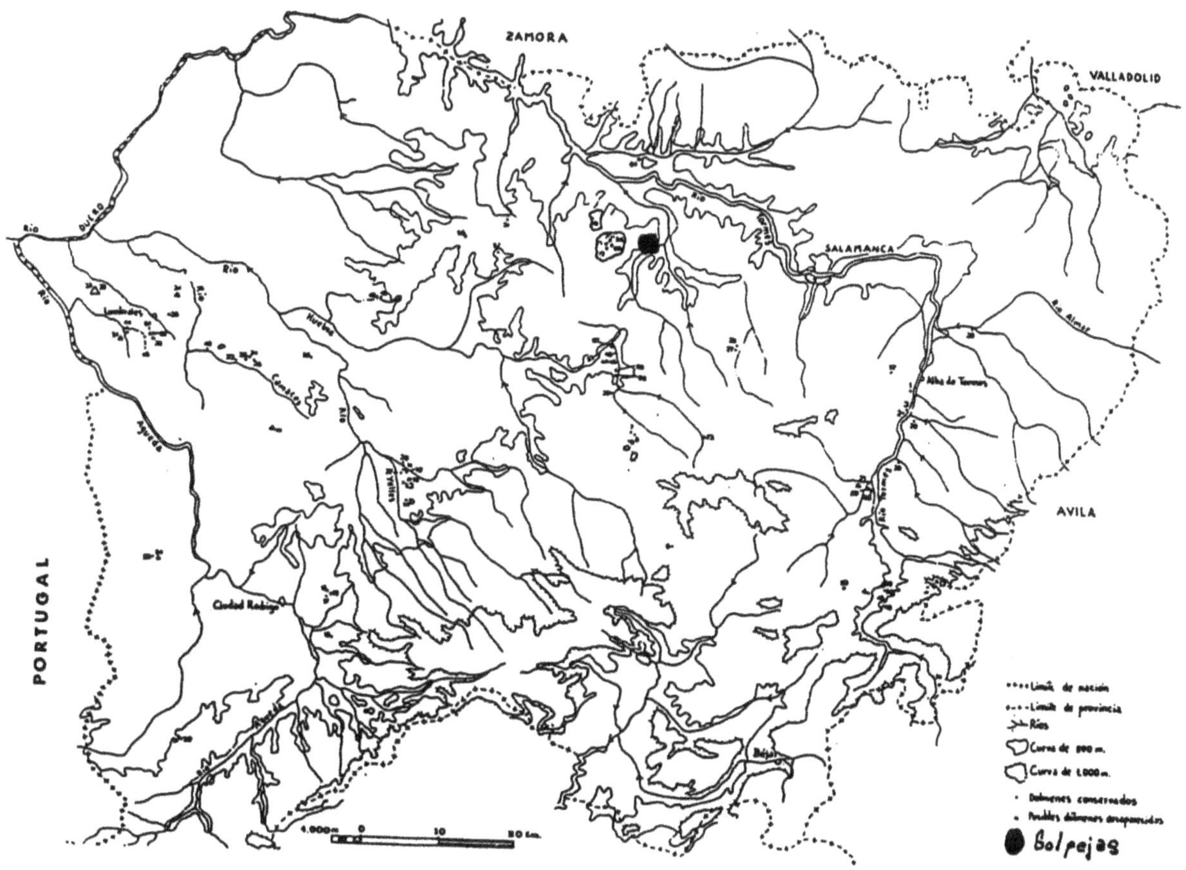

Fig 8.6 Geographic Location of the Megalithic Tombs known in the province of Salamanca
Delibes G. and Santonja M., 1986, 17-18

To the West and South of Golpejas, between five to eight kilometres are the remains of five dolmens (Delibes and Santonja, 1982, 17-19). Most of the Dolmens are within sight of the Golpejas mine area.

One of the Dolmens El Torrejon has been excavated by the Museum of Salamanca. I was allowed to examine the finds from the El Torrejon Dolmen excavations in the hope of perhaps finding that some of the tin/copper mineral might have been present amongst the finds.

I found one possible necklace bead which could consist of the Golpejas copper/tin mineral.

I now have received permission from the Salamanca Museum to bring the bead to the Institute of Archaeology for non destructive SEM surface examination.

Photo 20 Dolmen El Torrejon, 8 kilometers West of Mina Golpejas, showing different types of rocks used in construction, looking West

VISIT NO 17 **MINA BARRUECOPARDO**

MUNICIPALITY: BARRUECOPARDO

CO-ORDINATES: West: 6° 40' 30" - North: 41° 03' 15"
Provincial Map SALAMANCA 1:200.000 1984 2nd ed.

Photo 21 Mina Barruecopardo view of large open cast running North to South - looking South

The Barruecopardo Mine is now closed down. It probably closed 5-10 years ago from the look of the rusting machinery left lying around and the empty offices and store houses.

I spent half a day at the Barruecoparado mine trying to come to grips with the this very large mine area and to locate evidence of early metallurgy. No evidence of ancient metallurgy was found within the areas which I visited.

The Barruecopardo Mine consisted of several open cast pits (Photo 21) some of them extremely large running in a general North South direction.

The mine is no longer worked and most of the abandoned open cast pits are now partially filled with water.

Scheelite, $CaWO_4$ a tungsten ore and arsenopyrite were mined at Barruecopardo with tungsten being the main product. Small amounts of Tin (Cassiterite) were also found (Guzman, 1983, 55-7). The mine area is large. A proper archaeometallurgical surface survey would take weeks and would probably not be worth the time and effort necessary.

VISIT NO 18 **SITE OF LES MERCHANAS - RIO CAMACES**

MUNICIPALITY: LUMBRALES

CO-ORDINATES: WEST: 6° 40' 00" - NORTH: 40° 12' 55"
Provincial Map SALAMANCA 1:200.000 1984 2nd ed

Photo 22 Site of Les Merchanas - Rio Camaces - looking to South West

This location has been written up as being a military centre from which the Romans controlled the tin mining of the region in the 3rd century AD (Edmondson, 1987, 230). I visited this site therefore with high hopes of finding some slag samples or metallurgical working remains.

I found the supposed site of Les Merchanas about six kilometres to the North/East of Lumbrales. The site is named and marked on the Salamanca Provincial Map. The only visible remains at the site were two sets of ruins one the probable remains of a Roman Dam and the other ruins of a water mill with a final building phase most probably dateable to the 18th or 19th century (Photo 22). No significant amount of surface pottery was found at the Les Merchanas site and no metallurgical remains or slag.

It is possible that I visited the wrong site or that Edmondson (1987) has mis-interpreted his sources when writing about the Les Merchanas site and got his name and location confused.

Two people whom I met during my visit to the site said that it was not an area where the Romans had been known to be. The complete lack of any significant surface pottery seems to bear this out.

All in all, it was not a very useful visit. From information that I gathered later the old Roman area is located near or around the old train station in Lumbrales, which is near a small modern tin mine still in production.

I also made a quick visit to this area, but came up with no further evidence. It was not possible to gain entry to the working tin mine area.

VISIT NO 19 **MINA INESPERADA**

MUNICIPALITY: BARQUILLA

CO-ORDINATES: WEST: 6° 44' 30" - NORTH: 40° 41' 10"
Provincial Map SALAMANCA 1:200.000 1984 2nd ed.

Photo 23 Mina Inseperada - Barquilla, general view of Mine area - looking West

The only information that I had on the Mina Inesperada is that it was listed as a tin producing mine (Guzman, 1983, 47). The mine was owned and worked by a German Mining Company this century. The mine is no longer in production and looks like it has been abandoned for some time.

The mine area itself seems to have been a small operation consisting of a underground shaft and gallery mining system and accompanying surface waste heaps. The mineralization consisted of quartz, cassiterite, amblygonite and sulphides enclosed in quartz veins running in a Northerly direction (Junta De Castilla Y Leon, 1988, 84-6). The quartz veins are sitting in intrusive granites surrounded mostly by a complex of metamorphic slate (Grauvaquico).

I found no pottery, no ancient mining evidence or any archaeolo-metallurgical remains at this site. The ores were most probably located too deep to be worked in prehistory.

VISIT NO 20 **TORRE ROMANA CENTUMCELLAS**

PROVINCE: BEIRA BAXIA

MUNICIPALITY: BELMONTE

CO-ORDINATES: WEST: 7º 20' 20" - NORTH: 40º 22' 40"
BELMONTE no 225 1:25.000 SERIE M888 Edicao 1970

Photo 24 Torre Romana Centumcellas, 1992 - looking South West
the area by the car is where the surface tin slag was collected

The site of Torre Romana Centumcellas remains to date the one site from all my survey work in Iberia where I have encountered tin slag both surface and excavated.

The site of Torre Romana is located on the boundaries of the Portuguese provinces of Beira Alta and Beira Baxia just within the Northern boundaries of Beira Baxia and about two kilometres directly North of the town of Belmonte and approximately 449 meters above sea level. The site is surrounded by vineyards on three sides and a surfaced road to the North/East. The Torre Romana has at present a clear uncultivated area around the tower varying from approximately five to fifty meters (Photo 24).

It is this adjacent clear area which is at present being excavated by the Portuguese Archaeological Department (start 1993 - 1997 proposed finished date, an annual excavation season of four weeks).

The Archaeological excavation of the site was started in the summer of 1993 by the Portuguese regional Roman archaeologist based in Coimbra, Maria Helena Simoes Frade. Maria Frade excavated again in 1994 and will continue to excavate annually for a further two-three year period or until all the surrounding uncultivated area has been excavated and planned. The site will then be fenced and protected.

The destruction of the Torre Romana took place some time in the late Third or early Fourth Century AD (Maria Frade, 1994 personal communication).

The height of the Torre Romana ruins and location give it a highly visible presence in the Zezere Valley region. Domerque thinks that the tower might possibly have been part of a Roman control management-organisation centre, for the very extensive alluvial tin operations which the Romans carried out in the Zezere River and the Macainhas River valley regions (Domerque, 1990, 208-10).

J. de Alarcao believes that the tower at Centum Celas which was used as a watch tower in the Middle Ages could have formed part of a much larger, now destroyed, Roman villa complex (Alarcao, 1988, 70).

I first visited the site of Torre Romana in 1992 and have revisited the site again in 1993 and 1994. I also plan future annual visits while the archaeological excavations continue.

During my first visit in 1992 I found ten small pieces of slag on the surface just directly to the North/East of the stone tower. At the time of collection (Photo 43, page 143). I did not know what metal production had produced the slag, but had hopes that the pieces might be tin slag. The Torre Romana site was not being excavated during my first visit.

In the winter of 1992-1993 I was able to start analysing the slag samples which I had collected in the summer of 1992 from Torre Romana Centumcellas. The analytical work was carried out in the Wolfson Laboratories at the Institute of Archaeology, University College London.

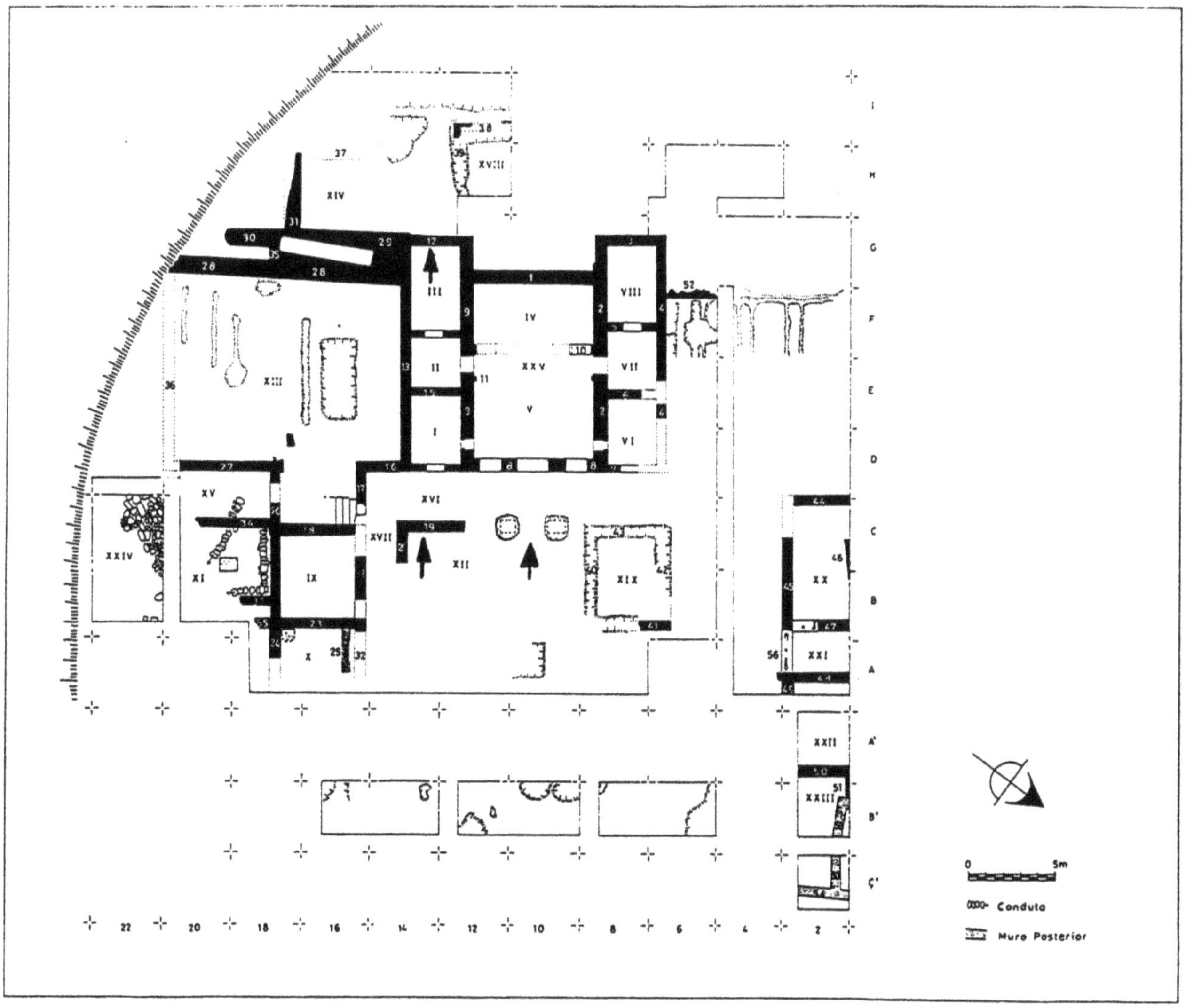

Fig 8.7 Site Plan Torre Romana Centumcellas-1994, arrows show locations of three pieces of excavated slag
Frade M. H. S., Directoria Regional do IPPAR de Coimbra, Portugal

I have done analytical work using both an SEM and Microprobe on the slag samples and the detailed results of this work are presented in the Analysis section of this report. The slag has proven to be a tin smelting slag (Sample No's 1 ,2 ,7 ,21 ,22 ,23 , page 144).

The geology of the Zezere valley and Belmonte region is part of the Central-Iberian zone of the Iberian Peninsula, consisting mostly of Hercynian granites formed around 280 million years ago and schists of the Schist-Graywacke Complex (Lourenco et al., 1992, 1). The area of Guarda, Belmonte and the Zezere Valley contains major and very wide spread deposits of alluvial and vein Cassiterite accompanied with Ilmenite $FeTiO_3$ and columbite.a combination of niobite $(Mn,Fe)Nb_2O_6$ and tantalite $(Mn,Fe)Ta_2O_6$ minerals.

It is probable that the tin slag was formed from smelting either vein or alluvial tin from which the (Ti) Titanium, (Ta) Tantalum and (Nb) Niobium had not been properly separated out during the ore preparation process. It is very probable that the Roman tin slag fragments from Torre Romana site were produced from the alluvial or vein ores of the Zezere Valley region.

During my second visit to Torre Romana in the summer of 1994 I met and talked with Maria Helena Simoes Frade who was just starting her excavations of the site.

Maria Helena Frade gave me the approximate dates of the site and also gave me three more pieces of small slag which she had just excavated (Sample No's 24, 25 ,26 ,27 , page 144-145). One (1) piece came from inside the foundation walls on the South side of the site and the second (2) and third (3) pieces came from the excavated area below where I had found most of my surface slag in 1992 (Fig 8.7).

Samples cut from the three pieces of excavated slag were analysed at the Institute of Archaeology and shown to have the exactly the same chemical make up as the 1992 collected surface slag from Torre Romana. There is no doubt that these three small slag fragments are from the same tin smelting process as the surface slag.

Maria Helena Frade also said in 1995 that she felt at this stage of her excavations the Torre Romana Centumcellas represents most probably the remains of a Roman Villa complex.

VISIT No 21 **MOUNTAIN OF FRAGUAS**

PROVINCE: BEIRA ALTA

MUNICIPALITY: GUARDA

CO-ORDINATES: WEST: 7° 13' 20" - NORTH: 40° 22' 40"
GUARDA no 214 1:25.000 SERIE M888 Edicao 1970

Photo 25 Looking South from the Mountain of Fraguas

The visit to the above site came about because of my lack of success in locating ancient alluvial tin working sites or remains on the Zezere and Macainhas Rivers in the Zezere Valley area around Belmonte. I was only able to find pieces of well washed and weathered pottery sherds, which were undateable, in the sand and gravel of the Zezere River near Belmonte.

I was told that an American Company had brought in large dredgers in the mid 70's and had worked out any remaining tin in the Macainhas and Zezere Rivers near Colmeal da Torre and Belmonte. It would seem most likely (if there had been any ancient remains of alluvial tin workings) that the big dredging operations of the mid 70's would certainly have destroyed most of the evidence.

I then enquired of some local people in the town of Macainhas if they knew of any areas close by where there might be old tin workings that were not alluvial. Two people told me that they knew of tin mining in the region of the mountain of Fraguas, about eight kilometres to the North East of Macainhas.

It was not easy to find the Mountain of Fraguas due to the poor maps I had in 1992, and the deep rutted unmarked dirt tracks which were the roads in the Frauguas area. I finally did locate the Mountain of Fraguas, but had no luck in sighting or locating any mining activity on or near the Mountain.

I decided to get out of the car and do a foot survey on the area of the mountain which looked the most promising for possible mining activities; this was the Mid-Northern slope of the mountain. I found no evidence of mining activity in this area, only a few small sherds of pottery which because of their small size and fragile nature were not really useful for dating. It is possible from the matrix that the pottery could have been Bronze Age or earlier (Luis Oosterbeek; personal communication, 1994), but if the pottery was old, it most probably only represented the remains of an early hunting camp. The supposed mining location information provided by the locals of Macainhas was not confirmed by my visit to the site.

VISIT NO 22 **THE MINE BY LAJES**

PROVINCE: BEIRA ALTA

ADMINISTRATIVE DISTRICT: GUARDA

CO-ORDINATES: WEST: 7° 42' 03".-.NORTH: 40° 28' 40"

BEIRAS FOLHA 4 1:200.000; CARTA DE PORTUGAL 1986
INSTUTUTO GEOGRAFICO E CADASTRAL

Photo 26 The Mine by Lajes - looking to South East

This mine site appears on a metal/mineral map of Portugal as an inactive wolframite cassiterite mine (Lourenco et al., 1992, 15-6). The Lajes mine area has been completely mined out and is now closed.

Geologically the Lajes site is located in the Portuguese part of the Central Iberian Zone. The Central Iberian Zone consists mostly of Hercynian formed granites approximately 250 million years old. The Lajes mine which is located on the edge of part of the Hercynian batholith was a lens deposit of cassiterite-bearing pegmatite (Dunning et al., 1989, 209). Pegmatite is a coarse grained igneous rock and is formed from the final portion of granite magma. Pegmatite normally contains quite high levels of metalliferous minerals. The metals are normally only present in trace amounts in the granite body from which the pegmatite originated (Bates and Jackson, 1990, 488-9).

It took a long time to find anyone in Lajes that knew anything about the mine location and operation. The supposed old mine workings were situated two kilometres slightly Southeast of Lages on the east side of the road to Lapa.

The mine was worked by a shallow strip mining technique, using a dredger, to a depth of approximately three metres. The remains of the mine workings today leave the impression that it was not a very large operation. It is possible that some of the mined land has been returned to agriculture use. The mine stopped working in 1977.

I spent about six hours doing a surface walking survey at the site and came up with a nothing of significance once again; no pottery and no evidence of any ancient mining or metallurgical workings. The site did seem right for possible ancient working due to the fact that the tin was located at no great depth. The nature of the modern shallow strip method of mining used at this site is very destructive and most probably would have destroyed or reburied any ancient evidence, if it had existed. I do not think that returning to the Lajes mine area in the future would be any more successful than my first visit.

VISIT NO 23 **THE GOIS MINE**

PROVINCE: BEIRA LITORAL

ADMINISTRATIVE DISTRICT: COIMBRA

CO-ORDINATES: WEST: 8° 05' 50" - NORTH: 40° 09' 40"

CARTA DE PORTUGAL 1986
INSTITUTO GEOGRAFICO E CADASTRDAL

Photo 27 View of the Gois area - looking to the North

The Gois mine is described as a metasomatic ore body which is surrounded by graphitic slate and consisting of very finely disseminated wolframite and cassiterite with cassiterite being the dominant mined mineral (Dunning et al., 1989, 210). There also appears quartz, iron-rich sphalerite, pyrite, and chalcopyrite.

I had no success in finding the mine. The people to whom I spoke at Gois did not know anything about a tin mine in the Gois area. I drove around the mountains directly to the North of Gois trying to locate mine spoil heaps but had no luck. The Gois tin mine must have been a very small affair or perhaps an underground mine. It must also have stoped working some time ago if the local people have no recollection of the mine.

I decided not to spend any more time looking for the mine, but I will go back for a future visit if meanwhile I am able to find more written information on the exact location and type of mine.

VISIT NO 24 **THE PANASQUEIRA MINE**

PROVINCE: BEIRA BAXIA

ADMINISTRATIVE DISTRICT: CASTELO BRANCO

CO-ORDINATES: WEST: 7° 45" 02" - NORTH: 40° 09' 50"

CARTA DE PORTUGAL 1986
INSTITUTO GEOGRAFICO E CADASTRDAL

Photo 28 View of the Panasqueira Mine and waste heaps - looking to the North East

The Panasqueira Mine area lies in a very high mountainous region, approximately 800-1000 meters above sea level, not far from the Rio Zezere river valley, about twenty four kilometers to the West of the town of Fundao and about eight kilometers to the West of the N 238 road (Photo 28). The mining operations at Panasqueira started in the mid-1930's.

The Panasqueira Mine lies near the southern edge of the Central Iberian Geological Zone and within an area of a Schist-Greywacke Complex.

The Panasqueira mine is foremost a wolframite mineral mine, with cassiterite as a secondary mined mineral (ratio 15/1 $(Fe,Mn)WO_4/SnO_2$) Argentiferous chalcopyrite is also mined for some copper and small amounts of gold (800ppm Ag) (Dunning et al., 1989, 210-12).

Most of the mineralization is found in the contact zone between a large buried granite body and its surrounding contact metamorphic aureole. The granite is considered to be post Hercynian due to its non distorted nature.

The mineral wolframite $W0_4$ and cassiterite $Sn0_2$ veins are centred around the apex and above the area of the buried granite; approximately at a depth of 620 meters and quickly die out towards the lower main granite body and the surrounding country rock.

The mineral veins have a varied mineralogy (arsenopyrite, pyrite, pyrrhotite, iron rich sphalerite and chalcopyrite). The mineral scheelite also appears at Panasqueira. The cassiterite is generally concentrated in the lower mineralised veins with the Wolframite appearing in the upper parts of the veins (Dunning et all, 1989, 210-12).

The Panasqueira mine operation was closed in 1984 due to falling wolframite and tin mineral prices on the world market. Only maintenance work is now being done at the mine above and below ground. The mine might reopen in the near future if wolframite mineral prices recover.

The modern Panasqueira mine was a very deep and extensive shaft and gallery operation and the surface area of the mine is mostly covered with modern waste heaps from crushing and concentration operations during its fifty years of operation. These waste heaps cover vast areas and could possibly cover ancient mine workings, had they existed.

I spent a full day at the Panasqueira mine area driving around, and on foot, in some areas that I thought might yield ancient metallurgical activity. The areas surveyed yield nothing in the way of evidence for ancient metallurgical operations either mining or smelting. One of the local people from Panasqueira (which is a dying mining town) said that a archaeological dig had been done in the area to the North West of the main shaft about three years ago and nothing important had been excavated.

Looking at the Panasqueira Mine area it would seem that the bulk of the cassiterite mineral is located at too great a depth to have been worked by ancient man. I will probably not return to this mine for any future survey work.

VISIT NO 25 **THE SANTA EULALIA MINE**

PROVINCE: ALTO ALENTEJO

ADMINISTRATIVE DISTRICT: PORTALEGRE

CO-ORDINATES: WEST: 7° 20' 15" - NORTH: 39° 1' 10"

ALTO ALENTENJO FOLHA 6 1:200.000
CARTA DE PORTUGAL 1970
INSTITUTO GEOGRAFICO E CADASTRAL

Photo 29 The Santa Eulalia Mine, general view - looking to North West showing shallow depth of modern workings

The Santa Eulalia mine is the most southerly of tin mine workings that I visited during my survey. The mine area is located in a North/West direction from the town of Santa Eulalia; from five to seven kilometres along both the north and south sides of the N 243 road.

Geologically the Santa Eulalia mine is located on or near the border area of the Southern edge of the Centro Iberian Zone and the Northern edge of the Ossa Morena Zone. See Fig 6.2, page 35.

The main ore body of the Santa Eulalia mine area is made up of metasomatized, greisenized granite which contains the minerals cassiterite and scheelite. The greisenization of the granite developed spaces within which small euhedral cassiterite and small allotriomorphic scheelite grains developed (Dunning et all, 1989, 209-10).

The Santa Eulalia mine was worked for cassiterite and scheelite. The mine worked for only five years (1976-1981). The depth of the mineral was only about one metre and consisted of a rich seam of tin/scheelite which was approximately ten centimetres in depth. This information is based on conversations with local people from Santa Eulalia in 1992.

The mine consisted of very large scale strip mining workings at a depth of only one to two metres. I was told by the local people that the total area of the workings was 890 hectares. It seems that most of the strip mining work was done by a mechanical digger or similar piece of equipment.

The mine workings consisted of three main area; two located to the north of N 243 and one to the south of the two mines to the north, the easterly one was called Herdade Do Miguel and the mine area to the West Baldio Do Monforte. The mine to the South of the road was also called Herdade Do Miguel and it was there that the main crushing /processing was carried out.

I was only able to do a surface survey of a few very small areas in the day and a half which I was at the mine. Most of the areas which I surveyed produced no evidence of ancient mining or metallurgy working.

It was the Baldio Do Monteforte mine area which produced the only finds.

It was in this area that I found surface pottery shards but no rims or bases, only pottery vessel body fragments, and possible Roman tegula (roof tiles) fragments. The collected surface pottery sherds were not good enough to use for dating.

Next to, and with the pottery and roof tiles I also found one round bun-shaped piece of slag approximately 8 cm in diameter and 2 cm thick (Drawing No 14). This supposed slag cake resembles the negative of the interior of a small smelting furnace bottom.

I analysed a mounted sample of the slag bun on the SEM and it turned out to resemble iron smithing slag. There was no tin present, just a very small amount of Titanium (0.21%) in one sample area (Sample No 18/B, page 150).

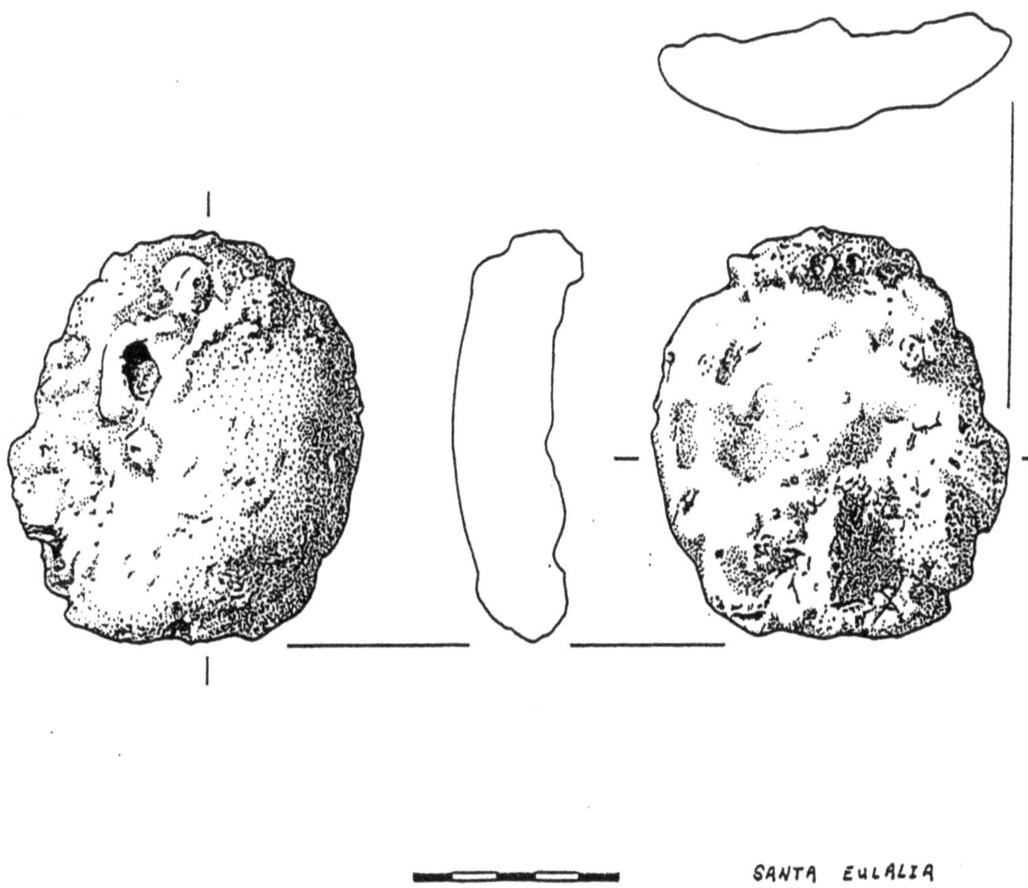

Drawing No 14 small slag bun, The Santa Eulalia Mine, scale units 1cm

Photo 30 Santa Eulalia Mine, general view showing Area where bun shaped slag and Roman tegula fragments found by 20 cm measurung rod

The Santa Eulalia Mine site is certainly an intriguing tin area due to the shallow nature of the cassiterite mineralization. The Eulalia tin deposits could have easily been available to ancient man. The problem in locating the evidence for ancient mining and metallurgy is the vast size of the mineral deposit and the destructive nature of the modern strip mining techniques.

I would like to visit the Santa Eulalia site again, but the problem remains, the gargantuan scale of the survey work and the time needed to complete such a large surface survey, which could only be done on foot.

I also collected what I thought might be some cassiterite mineral samples from the Santa Eulalia mine area. I analysed one of these samples (Sample No 10, page 150) in the SEM. The analytical results showed that 74% of tin oxide was present. Ancient metallurgists would have had no problems when working with acassiterite mineral which was as rich as this sample.

VISIT No 26 **RUINAS ROMANAS de SAO CUCUFATE**

VISIT No 27 **THE MINES OF AJUSTREL**

SOUTH PORTUGUESE ZONE-COPPER MINE AREAS

The above two visited sites were located in the copper mineral area within the South Portuguese Geological Zone. These visits were connected with my past studies of the South Iberian Copper Belt which covers both Spain and Portugal and will not be brought into my tin survey work.

VISIT NO 28 **LA MINA DE CASAS de DON ANTONIA**

PROVINCE: CACERES

MUNICIPALITY: CASAS de DON ANTONIA

CO-ORDINATES: WEST: 6° 15' 00" - NORTH: 39° 13' 40"
Provincial Map CACERES 1:200.000 1989 2a ed

Photo 31 La Mina de Casas de Don Antonia - looking North East

The mine at Casas de Don Antonia is listed as a cassiterite wolframite mine area (Ochoa, 1987, 104). The listing has two mines in the area, one wolframite and one cassiterite. The tin mine is a vein mine, the veins running in a N-S direction. The veins consist of quartz, cassiterite and mica encased in granite and hornfels rock.

It took me sometime to find the supposed tin mine area. I am still not sure if I found the tin mine or the wolframite mine. The mine which I did find was a very small affair which consisted of very small surface waste heaps and a nearby shaft (Photo 31).

The shaft was not too deep and did not seem to have any galleries running from it. I tend to believe that the galleries had been closed off and the shaft partly filled in when the mine ceased working.

I did a few hours of surface survey in the area of the shaft and waste heaps, but came up with no evidence of ancient activity pottery, mining remains or smelting activity. If I can get any more information about the exact location of the cassiterite minerals I might return.

VISIT NO 29 **MINAS VALDEFLOREZ**

PROVINCE: CACERES

MUNICIPALITY: CACERES

CO-ORDINATES: WEST: 6° 20' 10" - NORTH: 39° 27' 40"
HOJA 704 CACERES 1:50.000 1975 2a ed

Photo 32 Mina Valdeflorez general mine area view - looking to North East

The mine of ValdeFlores is listed as a tin mine, having produced evidence for being worked during the Roman period (Edmondson, 1987, 226). In his book "Two Industries in Roman Lusitania" Edmondson when referring to the antiquity of the Val de Flores Mine uses the following single reference source (Monteagudo L.,1954, 90; Caesaraugusta-4).

The ValdeFlores mine is located about three kilometres to the South/East of the centre of Caceres (Fig 8.8). In the bottom of a small valley (at 450 meters altitude) between the Sierra de Portanchito (644 meters) and the Cerro del Milano (566 meters) (Photo 32).

The mine works seem to have been abandoned some time ago and now consist of ruined mine buildings and some small surface waste heaps.

The modern mining method used at ValdeFlores was shaft and galleries to follow the irregular stockwork tin veins.

The final phase of the modern mine activity seem to have been the digging of exploration surface trenches mostly in North/South direction and approximately 2-3 meters deep running quite high up both sides of the valley.

Geologically the cassiterite of the ValdeFlores deposit is unique in that it has no known connection with a subjacent granite. It is assumed though that the ValdeFlores cassiterite veins must have some deep connection through the surrounding slate with either the great granite batholith of Cabeza de Arayato to the South/East or the Caceres granite batholith to the North/West.

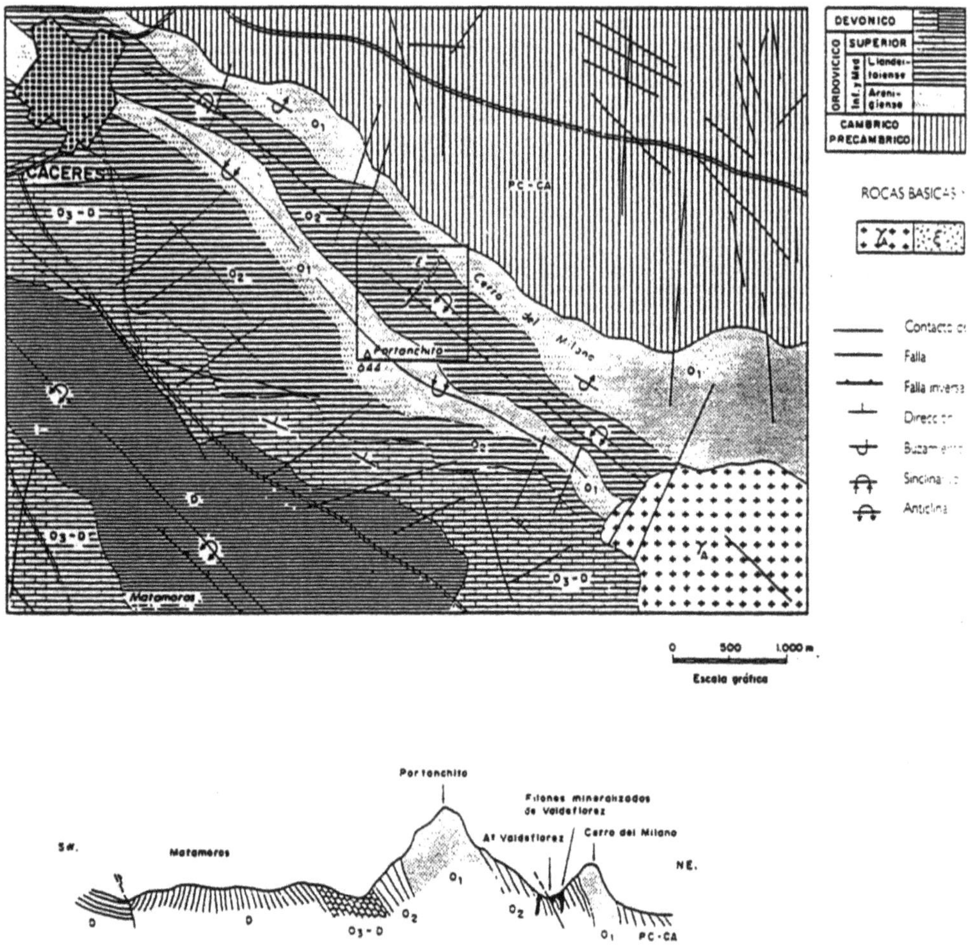

Fig 8.8 Placement and Characteristics of the Mineralized Veins of Ambylgonite- Cassiterite Valdeflorez (Caceres) Santos J.A. y Medina (1978); Molero, C.A.,

The main metals of the ValdeFlores mine are tin and lithium from the ores of Cassiterite $Sn0_2$ and Lepidolite $KLi_2Al[(F,0H)_2/Si_40_{10}]$ accompanied by scattered deposits of Turquoise, Scheelite and Apatite (Junta de Extremadura, 1993, 52).

The ValdeFlores tin veins are an irregular stock work and are encased in Ordovician (Lower Palaeozoic 500-435 million years old) slate and shale. The veins appear in two systems N 65^0 sub vertical and North 30^0West/40^0 South. The veins ranged in width from 10 cm to 120 cm with the average vein width being 20 cm. The minerals appearing with the cassiterite in the veins are quartz, amblygonite $(Li,Na)Al[(F,0H)/P0_4]$ and lepidolite $KLi_2Al[(F,0H)_2/Si_40_{10}]$ (Barrena and Isidro, 1987, 65-7).

The tin veins must have been visible in the small valley floor from a very early time and very easily accessible for trench mining as any rain most certainly would have washed away any covering of over-burden from a very early period.

The Val de Flores mine was quite easy to locate as there is a dirt track that goes directly from the East side of Caceres to the abandoned mine site. On my first visit to the Val de Flores Mine I found a single piece of slag in the road embankment. I kept looking for another hour hoping to find more slag, but had no success.

I cut the slag and sampled two different areas. The two mounted samples were analysed on the SEM at the Institute of Archaeology (Sample No 39 & Sample No 71, page 150). Unfortunately there were no traces of tin present in either of the slag samples.

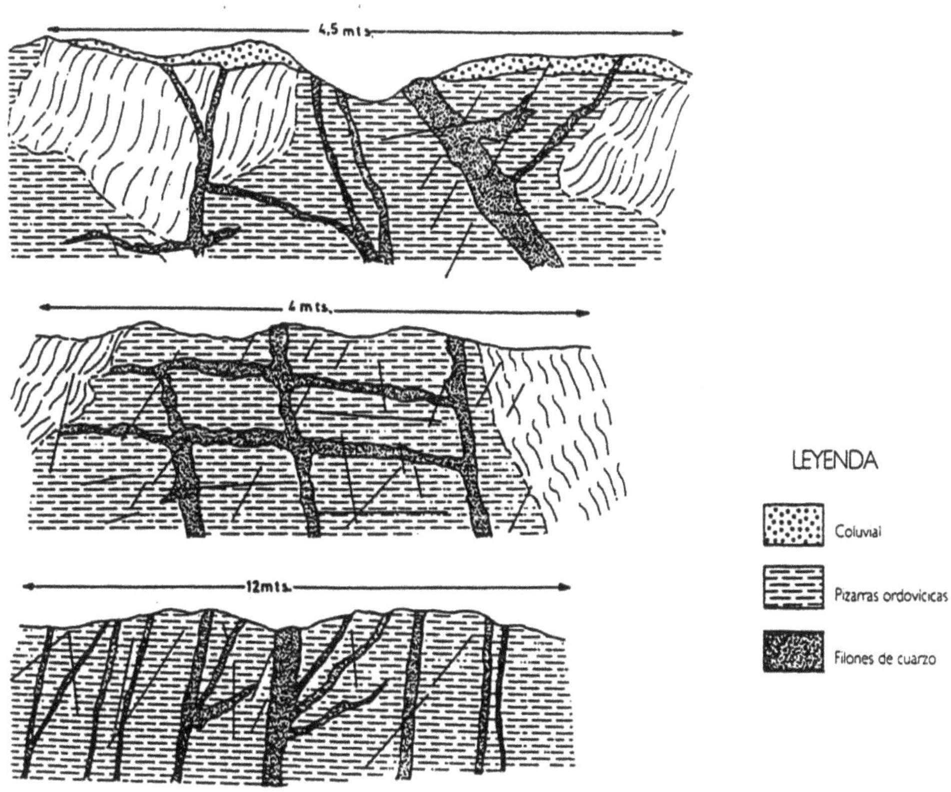

Fig 8.9 Geological Plan of the Valdeflorez veins, Santos, J.A. y Medina (1978) Molero, C.A., 1993, 98

Both of the samples are interpreted as iron slag with a small amount of Titanium present in most of the areas sampled averaging 0.40%-0.45%. This was again a disappointment: tin slag has not to date been found at ValdeFlores.

ValdeFlores is a mine area that I quite firmly believe must have been worked in ancient times, but so far, except for the one piece of slag, no other signs of ancient metallurgy have yet appeared. The possibility that the modern mining of the area has destroyed or covered any evidence must be taken into account.

There is at least one known Late Bronze Age settlement close by the ValdeFlores mine, the settlement of El Riscoa on top of Sierra de Fuentes (West: 06^{0} 17' 10" North: 39^{0} 26' 00"), approximately 5.5 kilometres to the South/East of the ValdeFlores mine site. This site has been excavated 1989 by Alonso Rodrigvez Diaz of Caceres University. He did not find any signs of metallurgical activity either smelting or casting at the El Risco site (Diaz, personal communication, 1992).

I visited the ValdeFlores mine again in April/May 1995, spending six hours at the site still trying to find evidence of ancient mining or smelting. The only surface find was a very small piece of probable slagged furnace lining located in the area where I found the first piece of slag. No pottery was found.

I found during my second 1995 trip that the ValdeFlores Mine area had been investigated by the Mining firm of Tolsa S.A., between 1985-1988, with the objective of seeing if the remaining mineral content of the stock work veins was great enough to allowing the re-opening of the mine. Forty two bore holes were sunk by Tolsa S.A. and all the old underground working galleries were opened and explored and the remaining mineral veins sampled and mapped (unpublished Tolsa S.A. exploration report, Servico de Minas, Merida).

Photo 33 Mina Valdeflorez, area by 20 cm measuring rod where slag found - view to West

The remaining underground stock work tin veins are quite small ranging from 1 cm to 20 cm in width. It seems that the answer of the investigation was negative and that all the original large veins of tin had been mined out in antiquity or during modern mining operations in the 19th-20th century (Junta de Extremadura, 1993, 52-6).

I will return to ValdeFlores for further surface survey work, as I still have a firm belief that the surface and underground vein tin deposits of ValdeFlores were worked in antiquity.

VISIT NO 30 **AREA CATRAIA DO BURACO**

PROVINCE: BEIRA BAIXA

MUNICIPALITY: BELMOMTE

CO-ORDINATES: WEST: 7° 20' 20" - NORTH: 40° 22' 40"
BELMONTE no 225 1:25.000 1970

Photo 34 area Catraia do Buraco - view to West, site area in foreground to left of dirt track in background Torre Romana Centumcellas

This site was found while I was doing further survey work in the small hill area directly to the North of the Torre Romana Centumcellas. I was looking at a mine which was indicated on the geology map which I had for the Belmonte area map 21A. The map indicated that the mine was in a Nb, Sn, Ta, Ti mineral area. When I found the mine it was a medium seized open cast pit partially filled with water and looked like it had not been worked for some time. It was not a very large mining affair. I did about an hour of survey work around the open cast mine area and did not find any evidence of ancient mining.

On the way back down the hill to Torre Romana I came across the site which I now call Area Catraia Do Buraco which is the name given to this area on map 225.

The site is located on the South side of a East/West dirt track which runs about 100 meters to the North of a line of houses located just north of the Torre Romana site (Photo 34). In the area called Catraia do Buraco I found very fragmented surface pottery, crushing/grinding and hammering tools, a saddle quern, a small loom weight (Drawing No 15, page 110), and two small pieces of slag. In a further visit to the site during 1994, I found two more small pieces of surface slag.

The surface finds suggest this area was a small settlement or a mineral/metal working area or possibly a mixture of both.

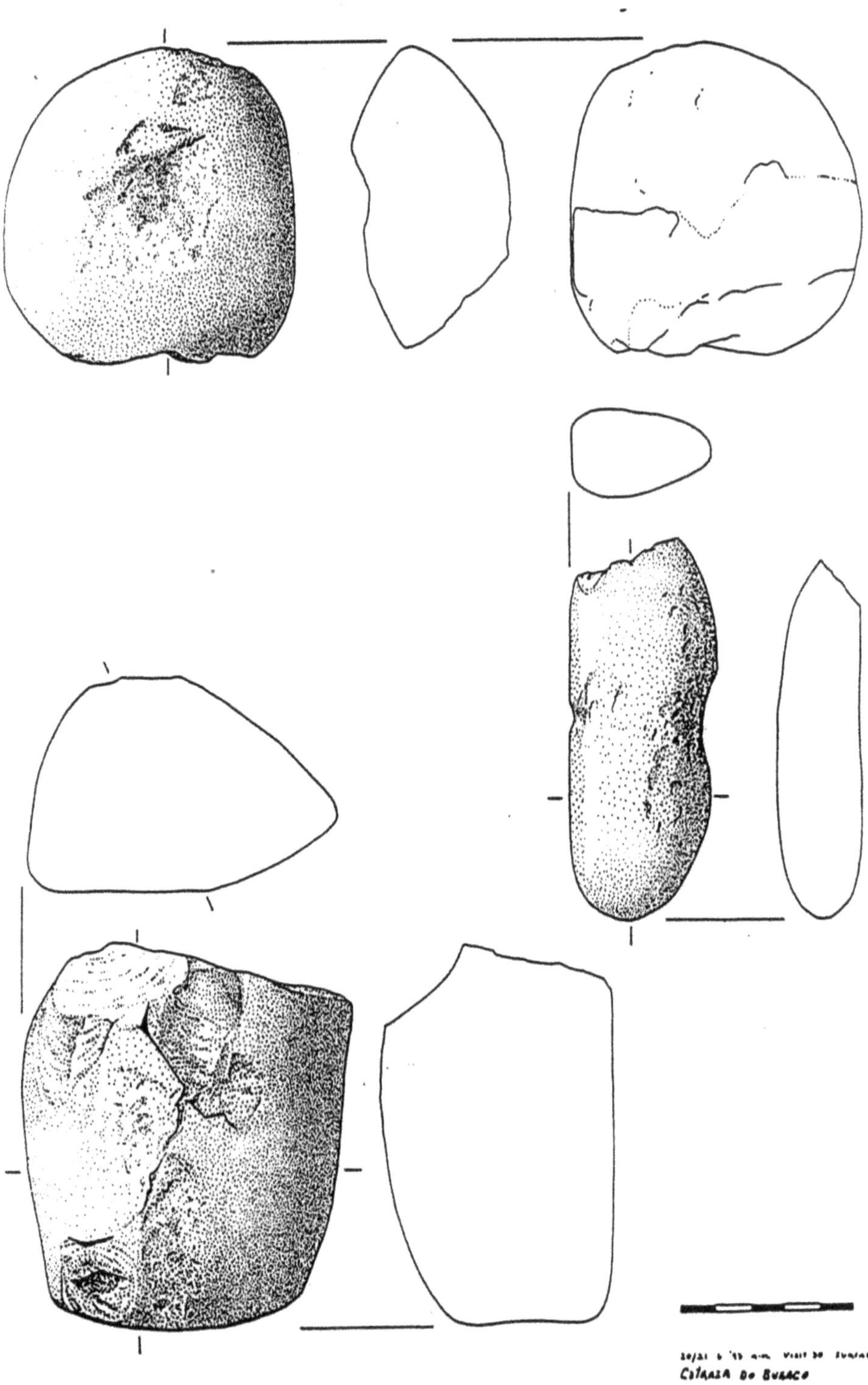

Drawing No 15 area Catraia do Buraco, worked stone fragments, scale units 1 cm

The fragments of pottery found at the site are so small and inconclusive that at present it is difficult to say exactly to which archaeological period the site belongs. The pottery was examined by Raquel Vilaca, personal communication (University of Coimbra, Portugal).

Pieces of the four small slag (Sample No's 28, 29, 30, 99, page 151) were cut mounted and analysed using SEM/EDS at the Institute of Archaeology. The slag analysis was interpereted as iron slag with the iron content varying from 28% to 54% depending on the area analysed. No Sn, Ta or Nb was detected by SEM/EDS in any of the analyses. On some areas there were small percentages of Ti. From the analysis results, it is not possible to call the slag from Area Catraia Do Buraco tin smelting slag.

Not far from the suposed settlement site, about 200 meters to the North/West, just to North of the same dirt track which runs by the Area Catraia do Buraco site, I came across a very well cut gallery which seemed to head due North. The gallery is cut into the local rock granite porphyry. The gallery measured at its entrance about 2 1/2 meters in height and 1/2 a meter in width. The lower 60 cm of the gallery was full of water. Part of the ceiling of the original entrance area of the gallery had collapsed for a distance of about 10-15 meters. I did not actually enter and explore the gallery.

Looking into the gallery from the entrance it seemed to get wider and then narrow down further on. I talked to some of the local people trying to find out if the gallery was a mine or a water supply. I received conflicting answers to my question, one person said that it was a tin mine and another said that it was a water supply gallery. Further future investigation inside the gallery might give a more positive answer as to its use.

I have a feeling that the gallery probably is part of a water supply system for the Torre Romana Centumcellas complex of buildings. Maria Helena Simoes Frade, the Portuguese archaeologist digging at Torre Romano, says that it is a water system gallery. She has also said to me that the general area to the North of Torre Romana Centumcellas seems to also have extensive ruins under the modern houses. It is within this area that the gallery is located.

VISIT NO 31 **SANTA MARIA MINE and SEARCH FOR MINERAL SAMPLES**

PROVINCE: BEIRA BAIXA

MUNICIPALITY: BELMONTE

CO-ORDINATES: WEST: 7° 19' 30" - NORTH: 40° 22' 15"
BELMONTE no 225 1:25.000 1970

After leaving the Torre Romana area I decided to see if I could find some of the numerous listed Sn, Nb, Ti, Ta mines located in the area around the Zezere Valley and Belmonte areas. The Carta Geologica de Portugal (21-A Sabugal) of the Zezere/Belmonte area shows the region blanketed with supposed Sn, Nb, Ti, Ta mine/mineral sites. I thought that it would be a good idea to collect some local mineral samples (if possible) for laboratory analysis. The samples if found would help to show the metal content of the tin ore.

I spent almost a whole afternoon walking around the Santa Maria hill area, but had no success in finding any appropiate mineral samples. I did locate one mine but it was very similar to the mine above Torre Romana which I had looked at the day before. It was a medium seized open cast pit; cut out of the North site of the Santa Maria Hill. I had no luck in finding decent mineral samples or any evidence of ancient mining or smelting at this minor mine site.

The next day I drove to other areas in the Zezere valley area around Belmonte which were listed on the Geological map as being mineralised Sn, Nb, Ti, Ta areas. These further surveyed areas did not yield any mineral samples.

It would seem that most of the Sn, Nb, Ti, Ta minerals of the Zezere Valley are very finely dispersed mineral particles within the granite porphyry or have been weathered and eroded out from the granite over the years and are present only in deep alluvial deposits in most of the local rivers.

It is thought that it was mostly the alluvial river tin which the Romans mined in the Zezere Valley (Alarcao, 1988, 79). I did collect some alluvial samples from the Zezere River by Belmonte and the Rio Inguias by the town of Inguias. The analytical results of these two samples were negative (no traces of tin were present). I hope that in future trips to the Zezere Valley I will be able to locate some local tin mineral samples for future analysis.

VISIT NO 32 **CUME 649 & 648 ABOVE QUINTA DE CHANDEIRINHA**

PROVINCE: BEIRA BAIXA

MUNICIPALITY: BELMONTE

CO-ORDINATES: WEST: 7° 21' 50" - NORTH: 40° 19' 45"
BELMONTE no 225 1:25.000 1970

Photo 35 Cume 649 & 648 above Quinta de Chandeirinha - view to South West

This site was brought to my attention by Antonio Augusto da Cunhz Margues who was working with Maria Helena Simoes Frade on the Torre Romana excavation. He told me that it was the best known Late Bronze Age settlement site in the Zezere River/Belmonte area which had not been covered over by any later settlement debris. This is one of the two sites that I visited during my survey that was not a mine area, but solely known as a settlement site.

The site is located approximately four kilometeres to the South/East of Belmonte on two small adjoining hill tops at the altitudes of 648 and 649 meters. The site sits on the South/Eastern end of the Sierra da Esperanca which is a very dominant mountain in the Zezere Valley approximately two kilometres in length.

It is a site that is a little difficult to get to as most of the old existing dirt tracks to the site have been closed off. The only way up is the road that goes through the small farm settlement of Quinta de Chandeirinha, located on the Eastern side of the Sierra do Esperanca.

It is a Late Bronze Age site typical for this area of the Western side of Mid-Iberia.

The settlement site is in a very prominent position and offers a magnificent panoramic vista of the Zezere and Macainhas River valleys (Photo 35).

There are still traces of a fortification or retaining wall on the North/Eastern side of the settlement. Someone quite recently has bulldozed a dirt track through the wall and I have a feeling that perhaps in the near future someone is planning to build a house right on top of the Bronze Age Site.

The site had plenty of surface pottery covering quite a large area. Stone querns were also scattered about. I spent the greater part of a day at the site. I felt that a settlement site which was located in the middle of a major tin mineral area just might yield some surface slag or other evidence of metallurgical activities.

I surveyed what I considered to be most of the site going by pottery coverage. No slag appeared even in areas which I surveyed on my hands and knees for considerable periods of time. If there is slag and evidence of metallurgical activities at this site the evidence will have to wait for a proper archaeological excavation unless the site is completely destroyed before then.

VISITS NO - 33, No - 34, No - 35

ZEZERE VALLEY: BELMONTE REGION

The surveyed areas of Visit's No 33, No 34 and No 35 should properly not have been given separate Visit numbers and in fact should be considered as a continuation of Visit No 31.

The three visits were a further searching for local tin mineral within the Zezere, Macainhas and Inguias River areas.

The three visits were made to locations which Antonio Augusto da Cunhz Margues (resident local archaeologist, Belmonte) said might have surface slag evidence.

None of the visited sites produced any surface slag or tin mineral samples.

VISIT NO 36 **CASTELO-BELMONTE**

PROVINCE: BEIRA BAIXA

MUNICIPALITY: BELMONTE

CO-ORDINATES: WEST: 7° 19' 50" - NORTH: 40° 21' 20"
BELMONTE no 225 1:25.000 1970

Photo 36 Castelo-Belmonte, area to North of castle keep where some slag fragments found - view to South West

The Medieval Castle area of the town of Belmonte was another location mentioned by Antonio Augusto da Cunhz Margues (resident local archaeologist,Belmonte, personal communication, 1994) as having surface and excavated slag. He said that he knew that slag had been found there during an archaeological excavation which had taken place within the Castle walls. In this excavation only small amounts of Roman pottery had been found. No Bronze age pottery appeared during the excavation.

I did not find any slag inside the Castle walls but in the area directly to the North/East, just outside the walls, I did find some surface slag. I also found surface slag both on the North/West slope and on South/East slope of the hill on which the Castle was constructed (Photo 36).

I collected about 65 pieces of surface slag from around the Belmonte Castle area (Photo 46). It would be impossible to put any date on any of this surface slag as it came from a urban area that had been continuously inhabited since the Medieval period and perhaps even earlier.

I was hoping that at least some of the collected slag would prove to be tin slag as the town of Belmonte is right in the centre of a very rich cassiterite mineral region.

I cut, mounted and analysed on the SEM at the Institute of Archaeology twenty four samples of the surface slag from Belmonte with very disappointing results (Sample No's 31-36 and Sample No's 53-70, pages 152-154). No tin or even traces of tin were present in any of the analysed samples. Once again the results of the SEM analysis pointed to all the slag being the product of iron smelting or smithing. I had hoped that at least one or two pieces of the twenty four slag samples from Belmonte might contain some traces of tin.

The results of all the analyse of the Belmonte slag are presented in the Analysis chapter. It still does seem strange that in a tin mineral region as rich and as well know as the Zezere Valley that any real evidence of tin smelting is lacking, except for the surface and excavated Roman tin slag from the Torre Romana area. I shall keep looking for further evidence of tin slag on my next visit to the Zezere valley area in the late spring of 1995.

VISIT NO 37 POVADO DA MISERICORDIA

SERPA / PORTUGAL

This visit to a supposed copper smelting furnaces which Monge Soares of the Quiminca Icenlineti, Lisbon had found within the South Portuguese Pyrite Belt geological region.

Therefore this visit site will not be discussed in this dissertation on the tin survey.

VISIT NO 38 **MINA de San EXPEDITO**

PROVINCE: CACERES

MUNICIPALITY: CACERES

CO-ORDINATES: WEST: 6° 21' 50" - NORTH: 39° 23' 58"
HOJA 704 CACERES 1:50.000 1975 2A ed

Photo 37 Mina de San Expedito, trench mine cutting vein in foreground - view to North East

The Mina de San Expedito is located about eight kilometres to the South/East of Caceres at kilometres 222 on the N 630 road travelling in the direction of Merida. There are two mines located within a kilometre of each other in this area. The Mina de la Union is at about kilometre 221.5 on the N 630 and about a kilometre to the North/West of the Mina de Sn Expedito (which in some books is refered to as the El Trasquilion Mine). The kilometres 222 and 221.5 of the N-630 are the distance from Madrid. These kilometre references from Madrid are posted on all national roads and marked on some maps.

Both of the two mines seem to have been closed down for some time. There is no trace left of the Mina de la Union which is located on the North/Eastern edge of a granite cupola, [in a contact metamorphosed shale area which was injected with mineralised quartz carrying veins (Dunning et al., 1989, 137)] I could not even see exactly where the Union mine had been, even when according to my maps and references I was almost directly on top of it.

Just to the North/East of the Mina Expedito is a small dammed lake (probably built to store water for the mine washing operations). The minerals mined contained Sn, Nb, Ta and were composed of irregular stock work veins scattered through out the granite (Ochoa, 1987, 100). The mining process at Mina Expedito also included the working of secondary alluvial and eluvial deposits in the area (Dunning et al., 1989, 137).

The only remnants at the Mina de San Expedito, (closed down in 1984) are some small/medium trench mining cuts running North/East to South/West and some medium sized waste heaps below the dam area (Photo 37). It was not easy to identify the Expedito mine area at first due to the complete lack of modern mining evidence, no ruined buildings or old rusting mining machinery.

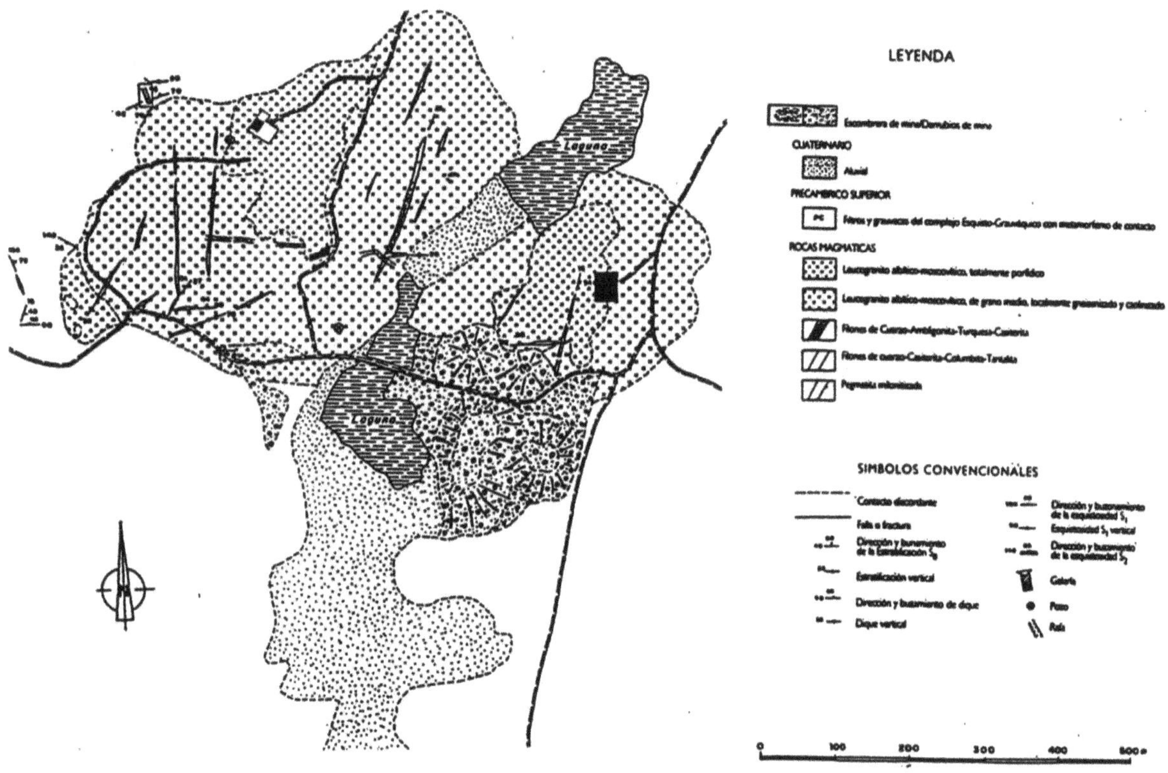

Fig 8.10 Geological Plan of the structure of El Transquilon, (Gallego M. 1992) Molero, C.A., 1993, 203

Geologically the mineralised area of the Mina de San Expedito Mine is a cupola granite extrusion, which is a surface projection from a larger and deeper granite body.

The Expedito Mine granite is greisenized (a process of hydrothermal alteration) and is made up of quartz, mica and topaz accompanied by tourmaline, fluorite, rutile cassiterite and wolframite (Fig 8.10). Parts of the granite body of Expedito had also been kaolinized (Molero, 1993, 94-5).

I spent half a day surveying the Expedito mine area and the only surface find was one piece of pottery. I found the piece of pottery about 100 meters to the South/West of the Casa del Transquilion.

Alonso Rodrigvez Diaz of the University of Caceres believes that the piece of pottery could represent a period earlier than the Late Bronze Age, but more pottery sherds need to be located before anything definite about dates can be decided about the Cassiterite Expedito site area.

The Mina de San Expedito is a interesting site due to the tin mineral being almost on the surface and therefore very easy to mine.

Unfortunately no surface evidence of ancient mining or smelting was found at the site. I will visit this site again and do further surface survey work in the near future.

VISIT NO 39 **EL MINA TERRUCO**

PROVINCE: CACERES

MUNICIPALITY: TRUJILLO

CO-ORDINATES: WEST: 2° 14' 00" - NORTH : 39° 28' 50"
HOJA 705 TRUJILLO 1:50.000 1a ed 1958

El Mina Terruco is listed as a vein and alluvial tin mine with the veins running in a North/East-South/West direction. The veins are composed of quartz and cassiterite and encased in a two mica granite rock (Ocha, 1987, 101).

I decided to visit this mine as it was listed in La Minera En Extremadura as being both an alluvial and a vein deposit Mine. It seemed to me that the quartz and cassiterite veins mineralization would have been on the surface and therefore would have been visible and available to ancient man.

Unfortunately I had no success in finding the former mine site at all. The abandoned mine site is supposed to lie 2 1/2 kilometres to the North of the N 521 on the East side of the Rio Magasea.

The dirt track along the Rio Magasea was blocked by a locked gate and a keep out sign. I then tried unsucessfully to drive to the mine site using some dirt tracks to the east of the site.

I did ask some local farmers in the area if they knew where the mine was or anything about it. They said that it been a shaft and gallery mine that had been abandoned for some years. The gallery is now blocked and the only thing that distinguishes the site are the ruins of a few old mine buildings. I will try to visit the El Terruco site again, it is most probable that I can only find it by walking to the supposed area where it is located.

VISIT NO 40 **MINA LA PERIZA - MONTANCHEZ**

PROVINCE: CACERES

MUNICIPALITY: MONTANCHEZ

CO-ORDINATES: WEST: 2° 28'00" - NORTH: 39° 12' 20'
HOJA 730 MONTANCHEZ 1:50.000 1a ed 1948

Photo 38 Mina La Periza, Montanchez - view to South West

I had been looking for the cassiterite mineral mines in the Montanchez area from the start of my survey work, but had no success in finding any references that would led me to the exact location of the abandoned Montanchez mines.

The minerals from the Montanchez area Mines are on display at the Geology Museum in Merdia. During my 1994 survey trip the custodian of the Geology Museum in Merdia, Tomas Porro Mayo was able to finally obtain and give to me, a photocopy of Sos Baynat's work on the Mines, Minerals and Geology of the Sierra De Montanchez (1981). This small publication of only ten pages helped me to locate the cassiterite mineral areas, which I had been looking for on the Sierra De Montanchez.

The cassiterite and wolframite of the Montanchez area are found only in thin quartz veins running in a North/East direction and originate from pegmatite formed granite rock (Sos Baynat, 1981, 5).

The Montanchez granite is located in one of the the southern boundary areas of a much larger stock granite field which is approximately 50 kilometres East/West and 20 kilometres North/South.

Almost all of the cassiterite mines on Sierra De Montanchez were worked by the digging of small open cast pits following the surface mineralised veins (Sos Baynat, 1981, 9).

All the mines in the Sierra De Montanchez have now been abandoned and have not been worked for some years.

Sos Baynat mentions ten mine areas in the Sierra De Montanchez which have been worked in the past for either cassiterite or wolframite.

I decided to try and find the La Periza mine area as it seems to have been the largest. It took a few hours to find what I believed to be the La Periza mine area. This mine area is located approximately 2 and 1/2 kilometres directly South of the town of Montanchez.

Once finding the mine area I was not able to do a surface survey due to the very dangerous circumstances which existed at the La Periza mine. The mine area consisted of numerous uncovered mine and mineral exploration pits; some quite deep. Almost all of the mine pits were covered or even concealed by very dense Mediterranean shrub and ground vegetation. Wandering around a mine area of this nature could be an extremely dangerous activity (Photo 38). Any survey work done must be done with the utmost caution, which I did not have the time for during my short visit.

I decided that I would make another visit to the La Periza Mine when I had more time and in a different season of the year, perhaps late summer early autumn, at this time of the year the shrub and ground vegetation should hopefully not be so dense.

I found no evidence of ancient metallurgy either on my walk to the La Periza mine or in any of the granite rock area surrounding the mine. to the North and West. I feel that this is a cassiterite mine site that could yield ancient mining remains if these remains have not been covered up or destroyed by later mining activities. The area of the town of Montanchez has yielded some evidence of human activity in the Late Bronze Age (Ignacio Pavon Soldevila, personal communication, University Caceres, 1994).

VISIT NO 41 **MINAS DONINOS de LEDESMA**

PROVINCE: SALAMANCA

MUNICIPALITY: DONINOS de LEDESMA

CO-ORDINATES: WEST: 6° 01' 35" - NORTH: 40° 59'35"
HOJA 477-1 SANDO 1:25.000 1a ed 1990

Photo 39 Minas Doninos de Ledesma - view to South West

The Minas Doninos de Ledesma mine area is listed as a small cassiterite/wolframite mine not far from the Golpejas Mine (Consejeria de Economina y Hacienda-Castilla Y Leon, 1988, 84). The abandoned mine area is located two kilometres to the South/West of the small village of Doninos de Ledesma (ten kilometres directly West of Golpejas).

Geologically the Minas Doninos de Ledesma is close to the edge of a very large granite consisting of biotite porphyry, a coarse grained granite which is intensely folded. The mine itself covers quite a large area and consists of numerous small mine workings of veins of quartz lying vertically in the granite. The thin quartz veins run in a general East/West direction and are mineralised with cassiterite, wolframite, pyrite, arsenopyrite, malachite and azurite (Consejeria de Economina y Hacienda-Castilla Y Leon, 1988; 84).

The above description makes Minas Doninos a very interesting mine area. Most of the quartz veins were probably visible on the surface and could easily have been seen and worked by ancient man. The minerals present at this mine cassiterite, pyrite, malachite and azurite could have all been used by ancient man.

My visit to this mine and the day spent doing a walking surface survey produced no surface pottery or any evidence of ancient mining or smelting activities. Once again what had seemed in the published literature like a good ancient mining and metallurgical site had produced no results. Perhaps the quartz mineralised veins were too thin and dispersed to have been noticed and used by ancient man.

Another visit to this site is necessary in the future.

VISIT NO 42 **MINA DE BERROCAL**

PROVINCE: BADAJOZ

MUNICIPALITY: MERDIA

CO-ORDINATES: WEST: 2° 38' 20" - NORTH: 38° 52' 20"
HOJA 777 MERDIA 1:50.000 1a ed 1941

Photo 40 Mina de Berrocal, northern side of granite mine area - view to South West

The Mina de Berrocal is another mine which I had heard about but could not find until, as mentioned earlier the custodian Tomas Porro Mayo of the Geology Museum in Merdia gave me a photocopy of a Sos Baynat book which contained more detailed information on the actual site location (Sos Baynat, 1965, 219 & 224). It seems that the Berrocal Mine has not worked for some time. It is not mentioned in any other more recent mineral/geological publications that I have on Extremadura.

The Berrocal Mine is located exactly at kilometre five on the road between Merida to Alange. This road follows the West bank of the Guadiana. The site from which I collected the surface finds (pottery and stone tools) (Drawing No's 16, 17, 18, pages 126-128) is located between the Guadiana and the East side of the road (Photo 41). Just to the south of the site are a complex of ruined mine buildings; last used in the 1950's and 1960's.

Most of the modern mining activity probably took place to the West of the road in the main granite approximately one kilometre by one kilometre in area.

Geologically the Berrocal mine area is one of three apophyses of granite, located along the Guadiana to the South of Merida. It is likely that these granites are connected to the batholith of Merdia.

To the North of Berrocal at kilometre three along the Merida-Alange road is the Fernandina-Cantarranas apophysis which is the largest of the three. This granite area runs under the Guadiana and surfaces directly opposite on the East side of the river. It is approximately five kilometres long and about one kilometre wide.

Photo 41 Mina de Berrocal, looking at settlement area by Guadiana River to left hand side of 1950's mine buildings - view to East

Two kilometres South of Berrocal also to the West of the Merida-Alange road is the third apophysis called Coscoja; the smallest of the three approximately 1/2 by 1/2 a kilometre in area.

The formation of the Berrocal granite is Hercynian in origin (approximately 280 million years old) and is aligned in a North/West direction. Coscoja, the most southern apophysis is a diorite. Berrocal and Fernandina-Cantarranas are normal granite, and granite with two micas. Located within the granite of Berrocal are veins which contain quartz and other minerals. These veins run in North/West and North/East directions. Pegmatites are also present and carry some of the mineralisation (Sos Baynat, 1965, 9 & 229 & 286).

The mineralisation present at Berrocal was mostly cassiterite, which appears in both veins and alluvial deposits, and wolframite which appears in vein deposits only. Also present in varying amounts is scheelite, chalcopyrite, cinnabar, pyrite, azurite and malachite. Mercury appears in liquid form in the alluvial soils in the area as does arquerite (Ag,Hg) with Hg (Sos Baynat, 1965, 305-8).

Isolated Gold nuggets were reported by the miners working at Berrocal Mine in the 1960's, but Sos Baynat was not able to confirm in his geological surveys of the area if gold did actually exist there.

Certainly the mineralogy of the Berrocal mine area is unique. Interestingly it contains various minerals that ancient man would have been able to recognise and work. Most of the minerals appeared in surface quartz veins in the granite which would have easily been visible to ancient man.

I have so far spent about two days doing surface survey at Berrocal. The small site area between the road and Guadiana River (North of the ruins of the mine crushing and washing buildings) has been the most fruitful area with surface pottery and stone tools.

The stone tools are varied and include a flat stone hammer, but could represent use from many different ancient periods. It is also impossible at this stage of the survey to tie the pottery area site to mining and metallurgy, the pottery could perhaps only represent a hunting camp or small early argiculture settlement site.

The main granite area to the East of the road is quite large and I have so far only spent a day surveying this area. I have probably only walked and looked at less than a quarter of the total area. Much more survey time will need to be spent in the Berrocal granite mine area (Photo 40).

Within the main Berrocal granite there does not seem to have been a single large mining area, rather a series of small and medium open cast cuts and trenches. Some of these were probably only mineral exploration trenches.

Also present are some areas which have been used as granite stone quarries, most probably for building stone use in Merdia. It is probable that some of the granite quarries are a secondary use after the area was mined out. So far very few fragments of surface pottery have been found within the mine area and the pottery of the mine area is not of a good enough quality to use for definite archaeological dating evidence.

The northern area of the granite produced a few probable surface stones close by pit/trench cut mines which might have been used as mining hammers. However another problem arises within the main mine granite area. Modern miners working at Berrocall in the 1960's would probably have expanded any ancient open cast pits and trenches they would have encountered.

Berrocal due to its mineral abundance and variation and location by the Guiadiana should be a prime site for evidence of ancient mining and metallurgy. The problem still remaining with the Berrocal site is finding archaeological and metallurgical surface evidence that will tie ancient man to the Berrocal granite mine area.

More survey work still needs to be done at this site. Also more effort needs to be made to learn more about the modern mining operations that took place at Berrocal.

I made three further visits to the Berrocal settlement and mine areas in April/May 1995. During my first visit I collected more surface pottery at the settlement site by the Guiadiana River.

In my second and third visit I concentrated more on the granite mine area trying to find evidence for ancient mining. These visits to the lower Northern area of the Berrocal mine site yielded two stone hammers from what could have been heaps of early mining waste debris (Drawing No's 19, 20, 21, pages 129-131). Pottery and flint was also found higher up in the Northern part of the granite mine area.

Alonso Rodriguez Diaz and Ignacio Pavon Soldevila (Departmento de Historia-Area De Prehistoria, Universidad de Extremadura, Caceres) have now had a good look at the surface pottery collected in 1994 at the Berrocall settlement site. They both, at this time believe that the pottery represents the transitional phase from Neolithic to Chalcolithic (Copper Age) within the mid Guadiana region from approximately 2000 BC to about 1800 BC.

The early pottery from the Berrocal settlement if connected to the mine area (unproven to date) could offer the possibly that gold and copper were being mined during the Copper Age at the Berrocal mine.

The settlement and granite mineralized area of Berrocal has now become a much more interesting site. The major problem facing this site is matching up the mine area with the settlement site. More survey work and time is definitely necessary on the Berrocal site.

I was also able to finally locate black and white photographs at the Merida Geological Museum that Sos Baynat had taken during his visits to the Berrocal Mine area in the 1950's and 1960's. They clearly show the granite mine areas being worked. I hope to be able to copy some these photographs during a visit to Merida in the late Autumn or early Spring of 1995/1996.

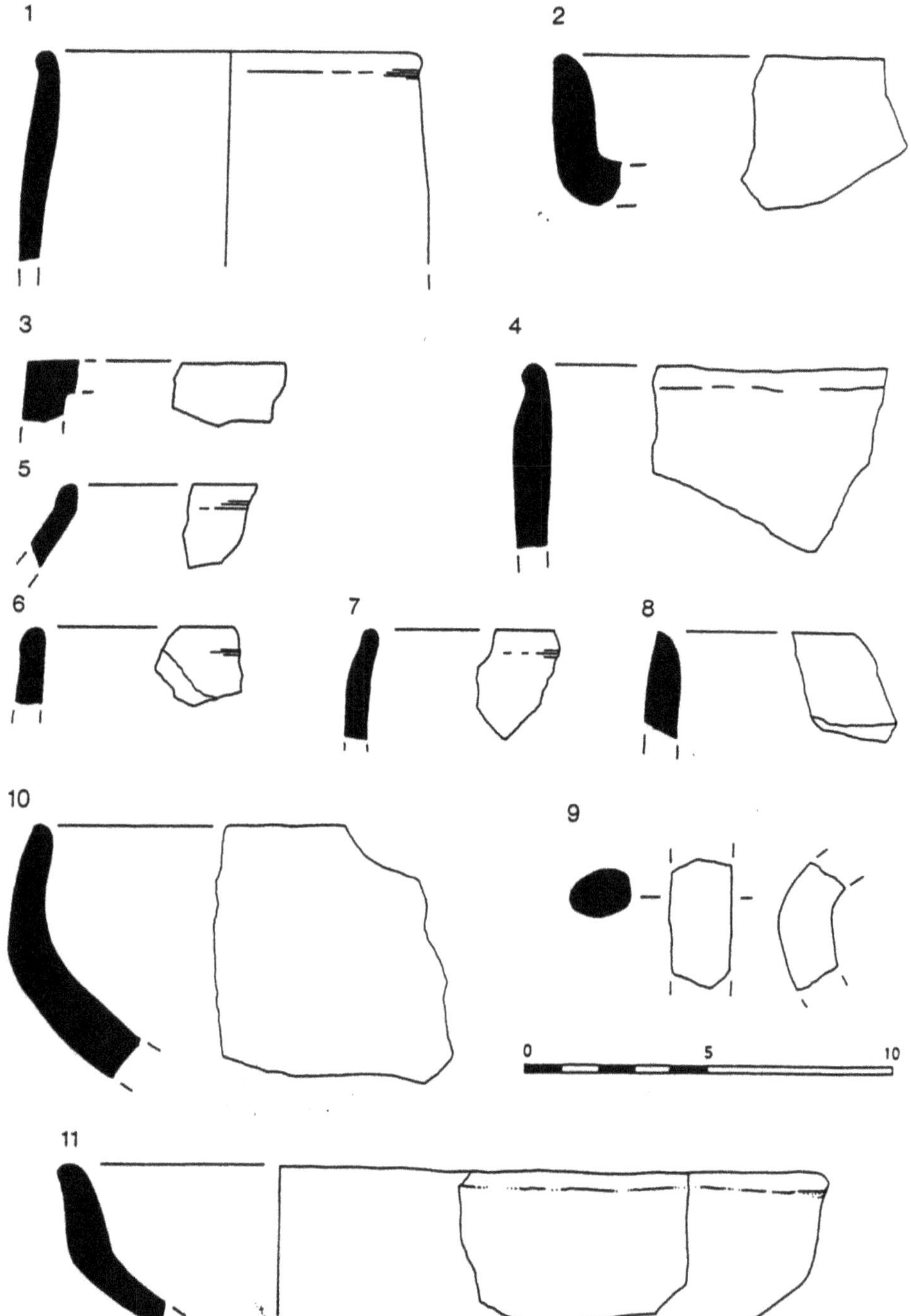

Drawing No 16 Mina de Berrocal, survey pottery, settlement site, scale units 1cm

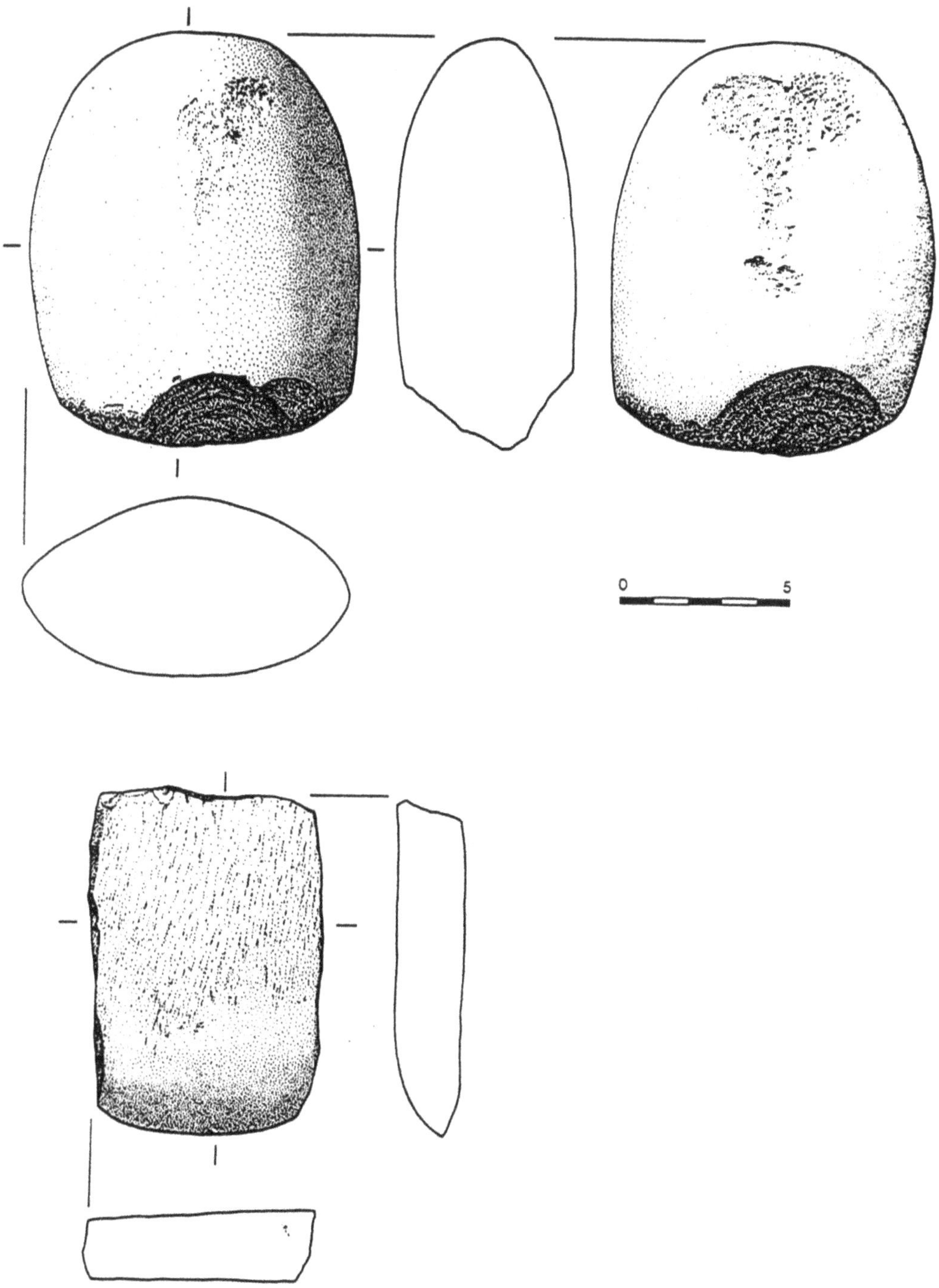

Drawing No 17 Mina de Berrocal, stone hammers, settlement site, scale units 1cm

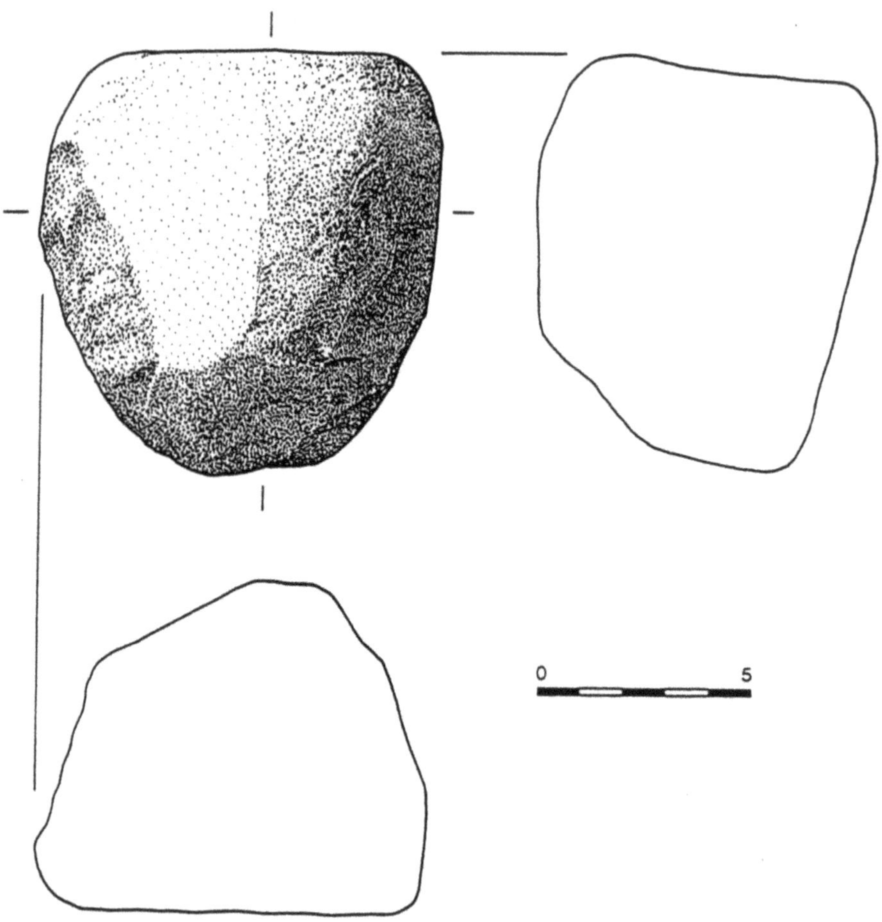

Drawing No 18 Mina de Berrocal, stone hammer, settlement site, scale units 1cm

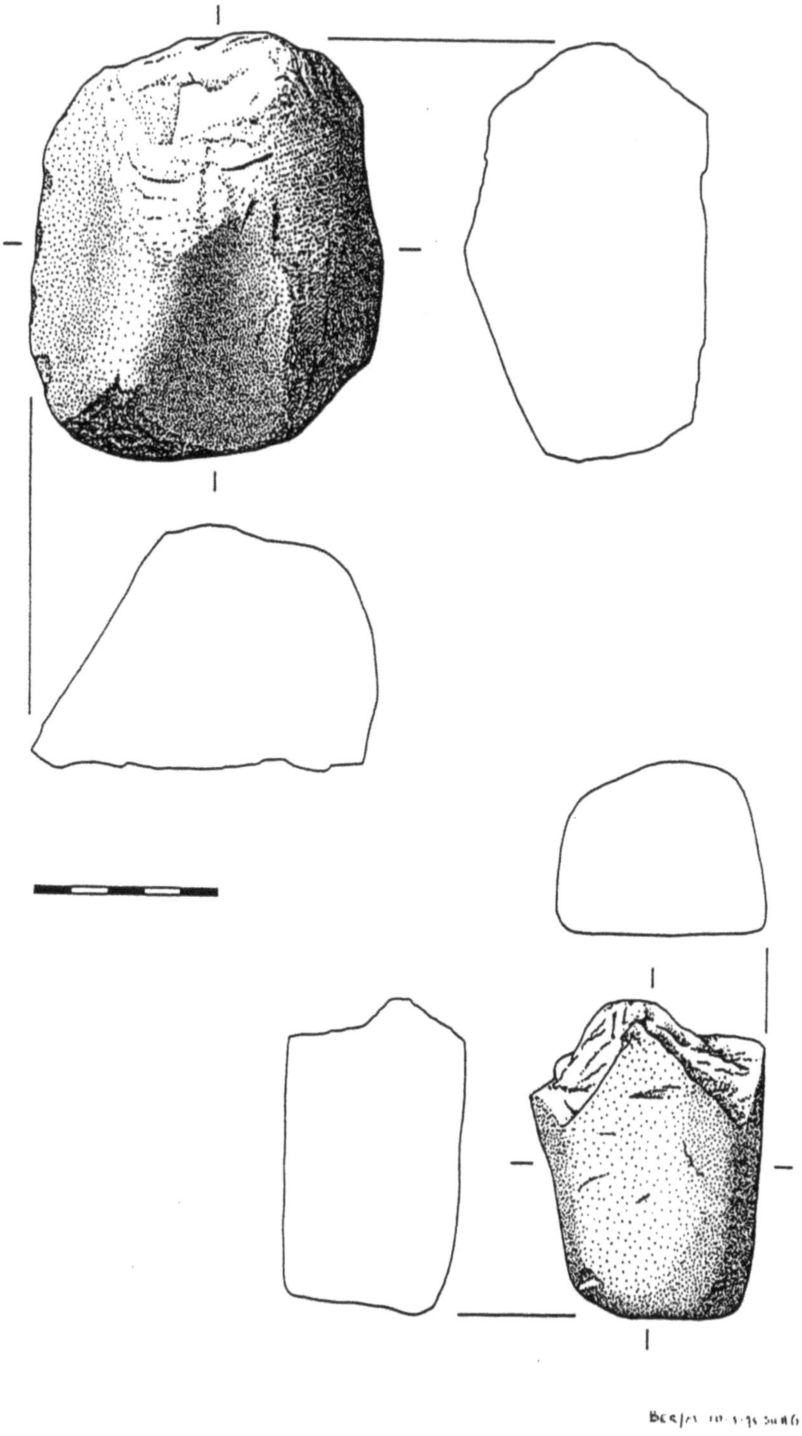

Drawing No 19 Mina de Berrocal, stone grinding/crushing fragments, scale unit 1cm

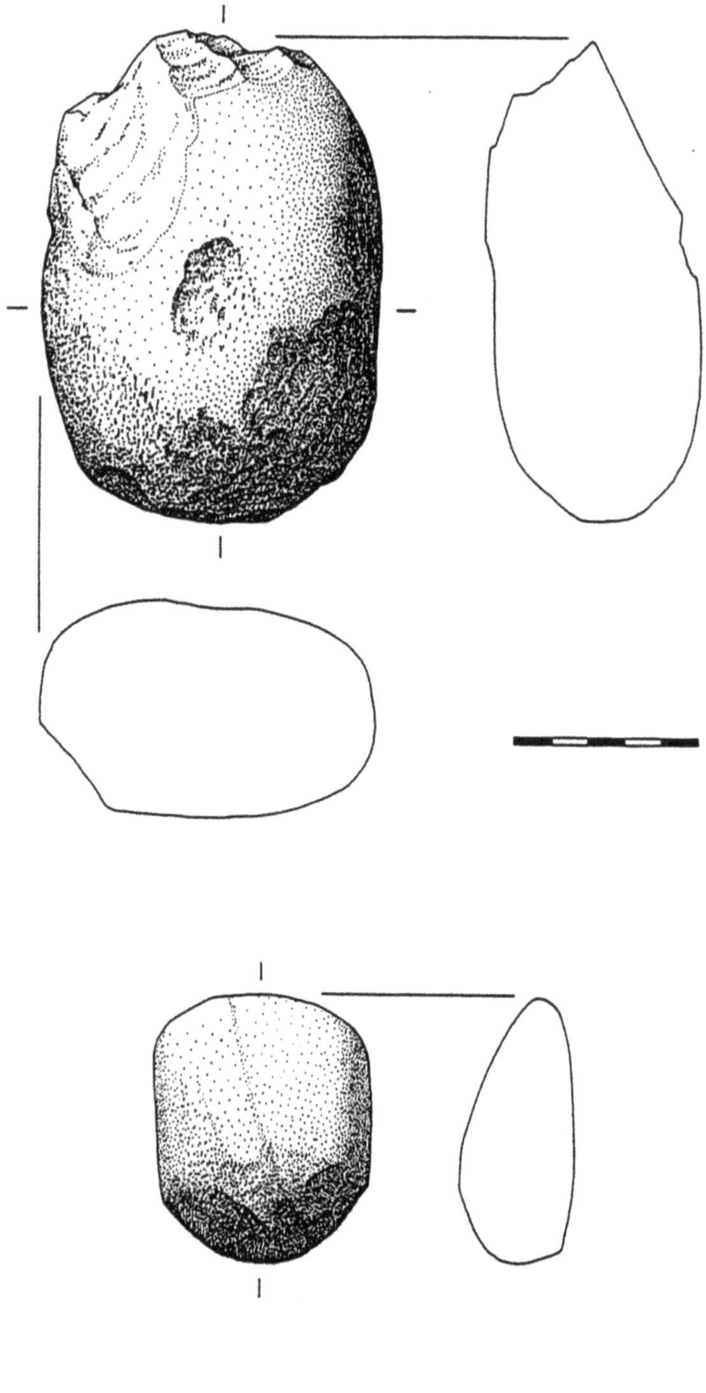

Drawing No 20 Mina de Berrocal, mine area, stone hammer and small grinding stone, scale units 1cm

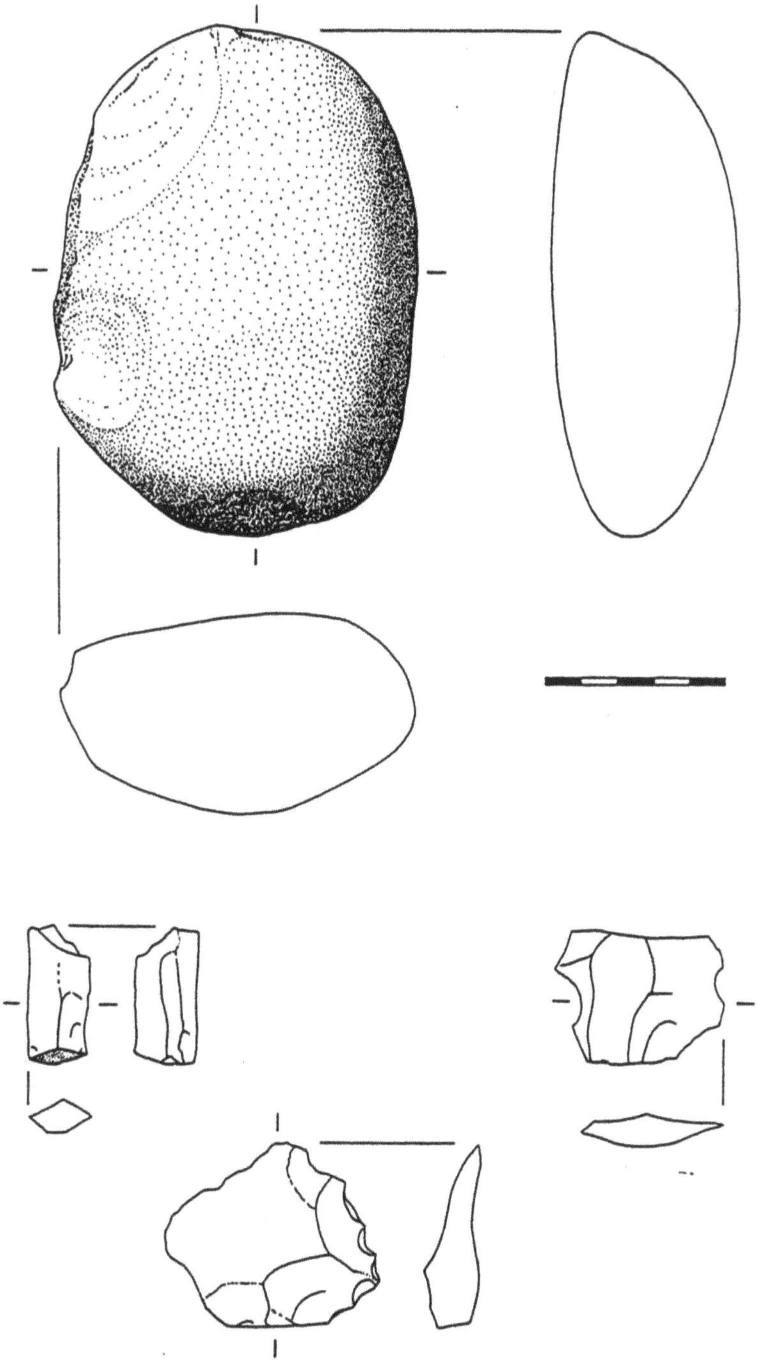

Drawing No 21 Mina de Berrocal, mine area, stone hammer and flint fragments, scale units 1cm

One hundred and three samples were collected in my field work. The majority of the samples came from ten survey sites. Not all of the collected samples have been analysed. The samples that had priority for laboratory work were slags as well as fragments of metal and selected minerals.

I had decided at the beginning of my field survey work to do a very superficial identification of metallurgical and mineral samples in the field and then to use laboratory analysis at the Institute of Archaeology London as the final arbiter for identification of the collected complex archaeometallurgical samples.

It is difficult to tell one metallic process extractive slag from another in the field unless upon breaking the slag visible metallic prills are present. If the metallic prills are not visible in a broken piece of slag then laboratory analysis is an absolute necessity for exact metallurgical process identification of the slag.

I also had a further problem during my survey work. The amount of slags which I found in the survey consisted of only a few small pieces [Tres Arroyos (3), Torre Romana Centumcellas (12), Santa Eulalia (1) and ValdeFlores (2)] except at Castle Belmonte in Portugal. Therefore I did not want to break the slag specimens open in the field as I felt this would be very wasteful of the small amount available for laboratory sampling. Cutting the samples with a diamond saw in the laboratory would provide a more efficient way for final slag identifications.

By including both field and laboratory work in my research my laboratory time was somewhat limited, due to the amount of time that I had already spent doing field work. I therefore chose Scanning Electron Microscopy with energy dispersive spectrometry (SEM/EDS) for microanalysis as a general analytical identification technique that was both quick and exact to the degree of identification that I needed for my field samples.

The mineral samples which I collected in the field using density and crystal identification (if possible) presented the same problems as the slag i.e. limited amount of material. For this reason SEM/EDS laboratory identification was also used to confirm or disprove my field judgements of the collected mineral samples and to provide a more precise estimate of the ores' purity.

By the second and third year of the survey my understanding and judgement of the minerals in the field had improved considerably, but laboratory analysis SEM/EDS was still needed to gauge the ore grade of the collected mineral samples.

The majority of my analytical laboratory work that was undertaken on my field collected survey samples was done using a quantitative Link AN 10 (Energy Dispersive Spectroscopy) EDS on an (Scanning Electron Microscope) SEM, Hitachi Model S-570. The laboratory work was done at the Wolfson Archaeological Sciences Laboratory, Institute of Archaeology, University College, London during 1993 and 1994.

The SEM and electron probe microanalysis (EPMA) are similar in that a beam of electrons interacts with the surface of the sample in the sample chamber, for imaging and spectroscopic microanalysis. The use of electrons offers much higher magnification capabilities (over 50,000 X) but this configuration for highest resolution gives shallow depth of focus; less than 10uM (Parkes, 1986, 185-6).

The image is produced on a SEM by collection of either backscattered electrons or the secondary electrons being emitted from the sample. Which system of collection and viewing is better depends on the material of the sample being scanned, the quality of the sample surface preparation and the type of information being sought.

The SEM/EDS offered a quick way of acquiring the basic qualitative elemental composition of each sample. The SEM does not detect trace elements, but as a primary analytical method it offered almost all the immediate information that was required.

Both the SEM and microprobe offer the opportunity to photograph the scanned sample.

Preparation of the samples included cutting, mounting, and a very brief optical microscope interpretation of mounted and polished metallurgical samples before carbon coating and final SEM analysis.

All the samples taken for analysis were cut from the collected metal artefacts, slags and mineral fragments using either a electric diamond circular saw or a jeweller's hand saw. These small cut samples were then mounted in clear "Epo-Kwick" Epoxy resin, which was allowed to harden for at least 8 hours; normally overnight.

The mounted sample resin blocks are 12 mm in diameter and vary from 10mm-15mm in height. The actual cut mounted sample material within the resin blocks ranged in size from 1-2 cm in the case of slag samples to 3 mm in size in the case of mineral samples, and in a few instances, sample sizes even smaller than 1 mm were also mounted and used for analysis.

The shape and size of the original collected samples slags, minerals and metallurgical fragments, determined how much of the sample would be cut for analysis. In some cases the original cut fragments from the samples seemed to contain no meaningful information, so a second or third section were cut off until either a visible metallic prill was exposed or the internal sample structure using a optical microscope had a definite visible structure.

Some of the collected field samples, metal fragments and slags had a deep layer of surface corrosion. Some metal samples' sub-surface areas had internal intergranular corrosion. This meant that quite a deep cut had to be made on these samples in order to try to sample areas that would not be contaminated by corrosion products, (which again was a difficult decision at times given the small samples).

All the circular mounted sample resin blocks were hand ground down and then polished.

The sample blocks were first ground down on varying grades of wet abrasive papers (120 grit to 1200 grit) to remove any surface distortion of the samples which might have occurred during cutting.

The samples were then further finely polished on two separate circular Metaserv Universal twin grinder/polishers, using polishing pads and diamond compounds as polishing agents.

The grades of the polishing compounds used were 30-um; 6-um, 1-um and 1/4-um. A liquid extender for the diamond pastes was also used during the polishing process.

The polished surface of all resin samples blocks were inscribed with a scriber to record top (T), bottom (B), left (L) and right (R) symbols next to the embedded cut/polished sample. This was done to ensure that the orientation of the original sample was maintained within the SEM specimen chamber or when examined with a optical microscope and that samples could always be viewed in the appropriate position.

The samples were then visually examined in detail under a stereo optical microscope. Ordinary reflected light and polarized light were used during these visual microscopic examinations.

The optical microscope was used to assess the general structure and to identify any inclusions of slag, oxides, or sulphides. Any porosity within the metal samples was noted. Identification of any metallurgical slag phases was based on optical properties.

The optical microscope was also used to locate and to possibly identify any metal prills present within the slag structures. Metallic prills found within the slag were then studied to try to determine visually, if possible, what metal was being smelted.

When metallic prills were found and if they were large enough a microhardness test was carried out using a small Vickers Optical Microscope testing unit. Hardness testing of metal does not give a completely accurate picture of the metal present, it is only a test of the metal's resistance to perforation or indenting (Brown, 1981, 89). The Vickers optical hardness test also helped by giving a preliminary quick identification of metal prills.

In the Vickers microhardness test, a time and pressure (30 seconds, 1 to 20 Kilograms) controlled pyramid shaped diamond indentation is made on the metal prill surface; then the measured length or width of the pyramid shaped indentation is located on a identification chart which assigns a set metal hardness value to the indentation lengths (Brown, 1981, 91).

Tin has a microhardness of 6-10 Hv, but measured values of metals are sometimes higher due to impurities being present.

The one factor which must really limit the use of the Vickers metal hardness testing on prills in slag is not absolutely knowing the thickness of the metal prill once it has been exposed in the cut polished sections. It is most probable that some of the indentions I made were being resisted more by the slag behind the sectioned prill rather than the actual metal making up the prill.

It was also at this stage of the laboratory sequence of work that some of the polished samples were photographed using a 35mm camera attached to the optical microscope. Tungsten colour transparency film was used for the photographs. The photography concentrated on prills and slag phase structure. All the sample blocks selected for examination in the SEM were carbon-coated in a vacuum coating unit.

Only one sample block at a time was inserted into the SEM/EDS specimen chamber. Two to five areas were analysed on each sample. It was necessary on some samples to analyse more than one area to check the heterogeneity of the metallographic phases. The average was about three areas analysed per sample. Samples which were unusual or showing a very diverse element content needed more area analysis. These samples had five to six areas analysed. All analyses were done on areas at X 800 magnification using a electron beam approximately <2 um in diameter.

The SEM/EDS linked computer programme was not used to normalise the final analytical results to one hundred percent. Instead the basic analytical information is presented exactly as it was recorded from the SEM computer program after ZAF corrections only. The final results in all samples are all presented in weight percent only. I did not round down from two decimal places on any of the EDS analysis.

The EDS does not detect oxygen, carbon and hydrogen. These elements were present in most if not all the samples analysed but oxygen and carbon varied considerably depending on the level of corrosion within the sample.

The difference to one hundred per cent on the final analysis sheets are assumed to represent the balance of undetected oxygen or hydrogen elements present in each sample. The percentage totals do fluctuate, and minor amounts of sulphur were present in some of the metal fragments.

The few bronze metal samples that have small amounts of arsenic present in the analysis (under 2%) could possibly represent the presence of earlier artefact fragments which had been mixed by resmelting. These earlier artefacts may have been broken up and remelted and recast as part of a larger bronze object production during the LBA.

Lead became generally used in the Late Bronze Age to make copper and bronze easier to cast, and in some regions reached levels between 5%-10% (Tylecote, 1992, 35).

The copper/tin samples are considered within the following general groups:

A low tin-copper mixture 6-12 % tin representing objects with a very good mix for working tools or pointed and edged weapons. These bronze artefacts would have been cast and then work hardened by further smithing using a hammering/annealing process.

A high tin-copper mixture 12-17% tin representing objects which would combine function with adornment (i.e. horse trappings, belt buckles) or very hard brittle work tools (i.e. awls, needles).

It is most likely that these high tin artefacts would have been cast only and not worked by a hammering/annealing process afterwards, due to the brittle nature that large amounts of tin bestow to bronze.

The high tin/copper bronze artefacts which were to be used for adornment would probably have been polished to a high reflective mirror-like surface finish. In my small amount of collected surface metal samples, no high tin-copper mixtures seem to have existed.

In the near future I plan to drill with (0.6mm and 0.8 mm drill bits) all of my metal samples. The drilled metallic turnings would then be used for Atomic Absorption Spectroscopy analysis, to detect minor trace elements present within the artefacts and produce better and more exact quantitative results.

As stated earlier it is also hoped in the near future to do further metallographic and microstructure analysis of some of my mounted metal samples. It was decided that the Atomic Absorption spectroscopy and metallography work were beyond the scope of the present PhD research.

Several samples (tin/copper mineral) from the Golpejas mine in the Spanish province of Salamanca and (tin slag) from Torre Romana Centumcellas in the Portuguese province of Beira Baixa proved to be very unusual. Therefore these two samples required further detailed analysis. To check and confirm the results of the primary (SEM/EDS) analysis.

The second analytical technique used on these two samples was a Electron Probe Micro-Analyser (EPMA). Two different Microprobes were used: one in the Wolfson Archaeological Science Laboratory (Jeol Superprobe) and one in the Birkbeck Geology Department (Jeol Superprobe 733 equipped with a Link AM 10,000/555 ZAF/4).

The Microprobe at the Institute of Archaeology was also used to scan and map the elements (Sn, Nb, Ti, Ta, Al, Si, Fe, and Mn) present in (Sample No 26) Roman tin slag from Torre Romana Centumcellas, Portugal. The microprobe scan/mapping process took almost 36 hours; due to the large number of elements being scanned. The results were stored on a memory disc as the sample was being scanned.

Later the stored results of the microprobe mapping/scan were brought up on the microprobe's coloured monitor screen. It was then possible using, a computer program, to produce a map of the slag consisting of any single, double or three scanned elements in any combination.

Single and various combinations of the mapped elements were then photographed off the monitor screen in colour negative and transparency film using a 35mm still camera. These photographic results offer a permanent record and a good starting point in the understanding of the make up and placement positions of the elements in the unique Roman tin slag.

The above microprobe scanning process was also used on Sample No 89 the copper/tin mineral collected at the Golpejas mine, Salamanca province, Spain.

XRD (X-ray diffraction) analysis was used on the tin/copper mineral sample from Golpejas as it was difficult to interpret the mineral composition and concentration using only the SEM/EDS analysis. XRD equipment in the Birkbeck Geology Department was used to analyse the tin/copper mineral sample from the Golpejas mine.

XRD (X-ray diffraction) analysis is primarily used in identifying the crystalline minerals which are present in geological samples. XRD analysis is ordinarily done using a powder diffraction method. A small sample (between 5-10mg) being analysed in this method is crushed to a fine powder. This small sample is then positioned in the centre of a cylindrical X-Ray camera. The camera monochromatic X ray beams are reflected (not scattered) back from the crystals in the powdered sample and captured on a detector or X-ray film negative. The exposed X ray film is then compared to other films of like crystals or XRD standard tables. These standard tables help to identify the various detected minerals. XRD analysis is able to detect minimum concentrations of crystal values as low as 1% to 10% of the whole (Parkes, 1986, 188-91).

EL CERRO DE SAN CRISTOBAL

LOGROSAN - Visit No 4 (Pages 51-66) Five samples were analysed. Fragments of three metal objects (a spearpoint, a chisel and a fibula) and two pieces of collected mineral samples.

(Sample No 44) Four areas were analysed on this spear point metal. All four analysis were almost identical and showed that the spear point metal fragment was a bronze mixture of approximately 10% tin with 90% copper and some small traces of sulphur and chlorine which possibly represent corrosion products.

(Sample No 97) Three areas were analysed on this metal chisel. All the areas were similar and represent a bronze mix of approximately 8-9% tin with 77% copper. On two of the analysed points small traces of sulphur were also detected.

(Sample No 98) Two areas were analysed on this possible fibula fragment. Both analysed areas were almost identical and represented a bronze of approximately 9% tin with 80% copper. Again on both areas there appeared small traces of sulphur.

(Sample No 102) Six areas were analysed on this mineral sample. The results showed that this mineral sample is a piece of Arsenopyrite. The arsenopyrite mineral (FeAsS) which had a high arsenic content 24%-40% in four of the six areas analysed.

(Sample No 103) Two areas were analysed on this mineral sample. This sample was cassiterite with 69% tin present.

EL CERRO DE		%	%	%	%	%	%	%	%	%	%	%	%
SAN CRISTOBAL		Cu	Sn	As	Pb	S	Fe	Si	Ca	Al	Cl	K	TOTAL
LOGROSAN													
Visit 4													
METAL	Area 1	75.92	11.51			0.49					0.22		88.15
Sample No 44	Area 2	87.62	11.70			0.59					0.23		100
SPEARPOINT	Area 3	89.95	11.62			0.55					0.15		102
20.2.93	Area 4	96.77	8.89			0.21					0.12		106
METAL													
Sample No 97	Area 1	77.26	9.11			0.26							86.64
CHISEL	Area 2	77.10	7.96			0.40							85.47
	Area 3	71.64	8.09										79.73
METAL													
Sample No 98	Area 1	83.66	8.71			0.59							92.97
FIBULA	Area 2	79.41	9.09			0.50							89.00
MINERAL	Area 1			24.05		0.76	20.22	0.39					45.43
Sample No 102	Area 2			5.22		0.17	5.10	21.53		16.12		6.73	55.09
ARSENOPYRITE	Area 3			40.08		10.81	31.25	0.56					82.75
	Area 4			29.51		3.25	22.71	4.49					60.15
	Area 5			0.71			0.53	41.14					42.42
	Area 6			36.71		8.37	29.11	0.68					74.86
MINERAL	Area 1		69.21										69.21
Sample No 103	Area 2		69.02										69.02
CASSITERITE													

Table No 1 SEM/EDS analysis, wt%, metals and minerals, El Cerro de San Cristobal, Spain

MINA TRES ARROYOS - Visit No 6 (Pages 69-70) Eight areas on three seperate surface collected slag samples were analysed. All three samples (No's 13, 14, and 15) analysed as slag were interpreted as having come from iron smelting. The iron content of all the points analysed was 48% and higher. Two analysed areas did have very small traces of Ti (0.1%). No Sn, Nb, or Ta were present in any of the analysis.

MINA DE TRES ARROYOS		% Sn	% Cu	% Nb	% Ti	% Ta	% Al	% Si	% Fe	% Mn	% S	% K	% P	% Ca	% Mg	% Na	% TOTAL
Visit 6																	
Sample No 13	Area 1						3.03	8.16	52.25	0.34		0.79		0.39			65.01
SLAG.																	
2.3.93																	
Sample No14	Area 1				0.11		3.38	9.54	47.26			0.39		0.42		1.13	62.24
SLAG	Area 2				0.11		3.20	10.97	47.07			0.54		0.59		1.62	64.14
2.3.95																	
Sample No 15	Area 1						4.51	12.01	48.25	0.22		1.01		0.36			66.39
SLAG																	
17.2.93																	
Sample No 15	Area 2						1.94		48.88			0.14		0.17	0.52		51.66
SLAG	Area 3						3.75		94.62			0.27		0.33	1.01		100
5.1.93	Area 4						2.16	17.80	78.34			0.25		0.38	1.04		99.99
	Area 5						1.34	11.07	48.73			0.15		0.23	0.65		62.20

Table No 2 SEM/EDS Analysis, wt%, slag, Mina de Tres Arroyos, Spain

LA LAPA MINE - Visit No 9 (Page 73) (Sample No 5) A single mineral sample was collected from the La Lapa mine. At the time of sample collection I was not sure if La Lapa was a tin or wolframite mine. The SEM analysis showed that in fact it was a wolframite mineral sample (43% W) with no minor trace of tin present.

LA LAPA MINA		% Sn	% W	% Nb	% Ti	% Ta	% Al	% Si	% Fe	% Mn	% S	% K	% P	% Ca	% Mg	% As	% TOTAL
Visit 9																	
Sample No 5	Area 1		45.23						11.48	2.43							59.15
MINERAL	Area 2		41.45						10.16	2.73							54.35
5.2.93	Area 3		42.95						10.70	1.97							55.63

Table No 3 SEM/EDS Analysis, wt%, mineral, La Lapa Mine, Spain

MINA DE DESQUITES - Visit No 10 (Pages 74-75) Two mineral samples were collected and analysed in the SEM from this survey site.

One mineral (Sample No 8) was a rich wolframite sample (55% W) and the other mineral (Sample No 9) probably could be considered a iron ore (30% Fe). I had hoped to find samples of cassiterite ore at this site; the analysis of the two collected minerals confirmed that I had not.

MINA DE DESQUITES		% Sn	% W	% Nb	% Ti	% Ta	% Al	% Si	% Fe	% Mn	% S	% K	% P	% Ca	% Mg	% As	% TOTAL
Visit 10																	
Sample No 8	Area 1		58.23						6.39	11.30	0.34						76.27
MINERAL	Area 2		57.40						6.92	10.23	0.21						74.77
9.2.93	Area 3		53.46						7.40	8.98	0.29						70.14
Sample No 9	Area 1								33.49		12.15		0.20			37.48	83.35
MINERAL	Area 2								29.87		0.90					34.80	75.33
5.2.93	Area 3								26.37		9.34		0.20			30.69	66.62

Table No 4 SEM/EDS Analysis, wt%, mineral, Mina de Desquites, Spain

MINA TELBA - Visit No 14 (Pages 80-82) (Sample No 4) The Sample collected at Telba was a cassiterite ore sample. It was analysed on the SEM solely to see the percentage of tin contained within the ore. Area analysis showed (70% tin).

MINA TELBA		% Sn	% W	% Nb	% Ti	% Ta	% Al	% Si	% Fe	% Mn	% S	% K	% P	% Ca	% Mg	% As	% TOTAL
Visit 14																	
Sample No 4	Area 1	75.83															75.83
MINERAL	Area 2	65.95															65.95
4.2.93	Area 3	71.65					1.75	1.92									75.32

Table No 5 SEM/EDS Analysis, wt%, mineral, Mina Telba, Spain

MINA GOLPEJAS - Visit No 16 (Pages 84-88)

A part of one of the Golpejas Mineral samples was ground down and analysed using XRD. The mineral consisted of three distinct different coloured constituents. The XRD analysis was done in the Geological Laboratory at Birkbeck College.

A mixture of white, green and dark areas consisted of:
Cassiterite SnO_2
Montebrasite $LiAl(PO_4)(OH)$
Wardite $NaAl_3(PO_4)_2(OH)_4.2H_2O$
Fluorapatite $Ca_5(PO_4)_3F$
Kinoite $Ca_2Cu_2Si_3O_{10}.2H_2O$

The mainly green material from the tin/copper mineral contained:
Montebrasite $LiAL(PO_4)(OH)$
Cassiterite SnO_2
Wardite $NaAl_3(Po_4)_2(OH)_4.2H_2O$
Kaolinite $Al_2Si_2O_5(OH)_4$

The mainly dark material from the tin/copper mineral contained:
Montebrasite $LiAl(PO_4)(OH)$
Cassiterite SnO_2
Wardite $NaAl_3(PO_4)_2(OH)_4.2H_2O$)

The mainly white material from the tin/copper mineral contained:
Montebrasite $LiAl(PO_4)(OH)$
Wardite $NaAL_3(PO_4)_2(OH)_4.2H_2O$

Table No 6 XRD analysis, mineral, Mina Golpejas, Spain

MINA GOLPEJAS - Visit No 16 (Pages 84-88) Forty-one areas were analysed SEM/EDS on six separate mineral samples. Five areas were also analysed on two mineral samples which had been heated with some charcoal (graphite) in small crucibles in a ceramic furnace (Sample 74 and Sample 82).

(Sample No 3) The three areas analysed on this sample showed it to be a piece of iron ore with a iron content of (36% - 47%).

(Sample No 6) Five areas were analysed on this sample. The mineral contained various percentages of Cu>(8.9% - 25.3%) and Sn>(20% - 34.5%) in all five areas plus trace percentages of Al>(0.4% - 0.6%), Si>(0.4% - 0.6%) and Fe>(1.3% - 2.3%), also present was P> in minor percentages (5.7% - 7.4%).

(Sample No 20) Six areas were analysed on this heterogeneous mineral sample. All areas had some percentages of Cu present. Sn was missing in three of the areas. The elements Al, Si, Fe, S, P, Ca, Na, and Cl were also present in varying percentages.

(Sample No 86) Eleven areas were analysed on this sample. The sample area consisted of two small native silver inclusions within the general mineral. No Sn appeared in any of the analysed areas, small traces of Cu (0.12% - 0.2%) did appear in all but two areas. Seven of the analysed areas contained Ag in very high percentages (47.0% - 82.9%) and another two areas contained Ag in small percentages (0.2% and 1.4%).

(Sample No 87) Six areas were analysed on this mineral sample. Sn (0.6% - 34.2%) and Cu (0.7% - 16.8%) appeared in all six areas analysed.

(Sample No 88) Ten areas were analysed on this sample and produced very diverse element analysis. Sn, Cu, and Ag, were all present in the sampled areas, but the percentages varied and only in Area 5 did all three appear together. Areas 1, 2, 7, and 8, are similar analyses of what are two native copper inclusions within the general tin/copper mineral matrix. Area 3 is a native silver inclusion. Areas 4, 9, and 10 could be considered standard analysis for the general Golpejas tin/copper mineral (Sn, Cu, Al, Si, Fe, P, Ca).

(Sample No 74) Smelted Mineral One area inside a metal prill was analysed on this carbon crucible smelted mineral sample. This area analysed as a naturally formed bronze with 6.25% Sn and 90.27% of copper.

(Sample No 82) Smelted Mineral Four areas, on this second piece of carbon crucible smelted mineral, located on small visible prills were analysed on the SEM. All four areas were very homogeneous in their element content, with a average of 35% Sn and 66% Cu and traces of Si, Fe, and S.

(Sample No's 6, No 20, No 86 and No 89) were analysed on the Microprobe at the Institute of Archaeology. The microprobe was used to check and confirm the SEM analysis, only graphs showing the elements present were produced. Sample No's 20, No 86 and No 89 confirmed that Tin and Copper were present throughout the mineral sample. Sample No 86 confirmed that Ag inclusions were present in the tin/copper mineral.

MINA GOLPEJAS		% Sn	% Cu	% Ag	% Br	% Al	% Si	% Fe	% Mn	% S	% I	% P	% Ca	% Na	% Cl	% TOTAL
Visit 16																
Sample No 3	Area 1					1.20	1.14	47.2								49.50
MINERAL	Area 2					0.94	8.01	36.9								39.50
9.2.93	Area 3					1.19	1.19	46.3								58.73
Sample No 6	Area 1	24.5	23.7			0.38	0.59	1.78				7.01				58.03
MINERAL	Area 2	34.6	8.95			0.71	0.46	2.31				6.53				53.57
28.1.93	Area 3	31.4	13.4			1.25	0.60	2.09				5.67				54.43
	Area 4	20.0	25.3			0.64	0.44	1.32				7.38				55.18
	Area 5	23.9	22.1			0.45	0.41	1.72				6.61				55.27
Sample No 20	Area 1		10.1			2.45	24.8			0.28		7.93	9.11		0.19	54.97
MINERAL	Area 2		6.52			0.62	2.42			0.07		0.55	4.83			15.04
17.2.93	Area 3		0.86			13.5				0.11		16.3	1.91	1.93		34.76
	Area 4	27.5	8.17			4.05	0.96	1.93		0.31		8.07	11.6			62.69
	Area 5	1.31	10.0			2.66	25.6	0.26		0.23		7.95	9.36		0.16	57.6
	Area 6	31.3	3.11			1.46	0.75	2.31		0.49		8.99	15.7		0.12	64.27
Sample No 86	Area 1		0.34	80.7	0.59						0.36					81.98
MINERAL	Area 2		1.79	82.9	0.82											85.71
22.12.93	Area 3			68.3												68.31
	Area 4		0.50	82.1	0.49						0.56					83.62
	Area 5		0.53	60.8	20.9						4.83					87.13
Sample No 86	Area 1	0.29	68.4													69.7
MINERAL	Area 2		0.26	47.0	9.01						1.88					58.52
7.1.94	Area 3		2.19	1.42		6.43						8.16		0.84		21.11
	Area 4		0.24	0.22	7.00	3.37						9.10	0.47	0.82		21.26
	Area 5				5.65	2.62						9.68	0.12	0.14		18.32
	Area 6		0.15		4.78	2.62						8.05	1.11	0.14		16.89
Sample No 87	Area 1	29.8	14.0			1.79	1.42	4.72		1.35		2.16				55.29
MINERAL	Area 2	27.2	16.5			0.59	1.06	5.22		1.20		1.20				53.01
6.1.94	Area 3	0.64	0.71			12.1		0.13		0.11		17.1		0.57		31.45
	Area 4	24.5	16.9			2.40	0.49	3.26		1.08		4.87				53.46
	Area 5	33.5	11.5			0.70	0.74	4.42		0.64		1.86				53.41
Sample No 88	Area 1		80.5	0.37		0.57	0.40					0.32				82.39
MINERAL	Area 2	0.31	89.8				0.68					0.42				91.37
6.1.94	Area 3			35.3		0.18	0.26				47.8					83.95
	Area 4	28.2	17.1			1.81	0.73	1.64				3.85				53.36
	Area 5	1.63	0.68	10.7		7.43	0.28				8.79	12.8				42.47
	Area 6		22.2	0.37		6.02	0.35	0.15				14.0	16.4		2.07	61.64
	Area 7	83.4			0.41	0.38					0.40	0.41			5.57	90.7
	Area 8		95.2				0.37									95.95
	Area 9	36.2	2.43			0.48	0.43	2.55				8.10	13.1			63.42
	Area 10	11.9	0.99			1.60	0.45	0.78				14.0	24.3			54.20

Table No 7 SEM/EDS analysis, Wt%, mineral, furnace smelted mineral; Mina Golpejas, Spain

MINA GOLPEJAS		% Sn	% Cu	% Ag	% Br	% Al	% Si	% Fe	% Mn	% S	% I	% P	% Ca	% Na	% Cl	% TOTAL
Visit 16																
Sample No 74																
MINERAL	Area 1	6.25	90.3							0.13						96.65
FURNACE																
SMELTED																
9.12.93																
Sample No 82	Area1	35.4	66.8				0.13	0.30		0.18						102.8
MINERAL	Area 2	35.9	67.0					0.32		0.24						103.5
FURNACE	Area 3	35.8	67.0				0.13			0.07						103.0
SMELTED	Area 4	36.1	67.1													103.5
16.12.93																

Table No 7 SEM/EDS analysis, Wt%, mineral, furnace smelted mineral; Mina Golpejas, Spain

The microprobe at the Institute of Archaeology was also used on Sample No 89 for the same scan/mapping procedure which I used on the Torre Romana slag (Sample No 26). The Tin/Copper mineral was scanned for seven elements Sn, Cu, Fe, Al, Si, Ca and P.

Coloured photographs were produced showing supposed single and mixed element compositions. The coloured Photo 42 below should be viewed more as a configuration map of a polished section of the Golpejas mineral than a absolute correct placement of the Sn within the sample.

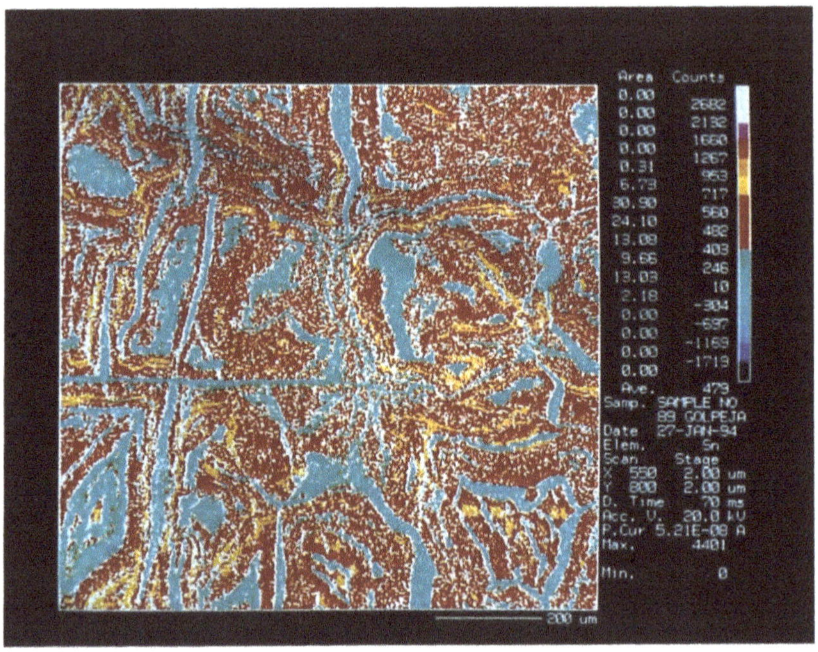

Photo 42 Coloured Microprobe Configuration Scan Golpejas Cu-Sn Mineral, Sample No 89

The Microprobe at the Geology Department of Birkbeck College was used to check the results of the Institute of Archaeology SEM and Microprobe work. Sample No 6, No 20, No 86 and No 88 were all analysed using a quantitative program.

The Microprobe results matched the analysis that had been done at the Institute of Archaeology in showing that native copper and silver inclusions were definately present within the general tin/copper mineral matrix. Though the same samples were done on the SEM and Microprobe some of the areas analysed were different locations within the samples.

MINA		%	%	%	%	%	%	%	%	%	%	%	%	%	%	%
GOLPEJAS		Sn	Cu	Ag	Ti	Al	Si	Fe	Mg	S	Br	P	Ca	Na	Cl	TOTAL
Visit16																
MICROPROBE																
BIRKBECK																
Sample No 6	Area 1	79.5														79.49
MINERAL	Area 3		5.85	79.4								2.10				87.38
31.1.94	Area 4	1.89	3.64	88.4				0.26	0.18			0.60				94.97
Sample No 20	Area 1	75.53	1.39													76.89
MINERAL																
31.1.94																
Sample No 86	Area 6			97.16					0.17	0.09						97.42
MINERAL	Area 7			82.10		5.32			0.24	0.40						88.06
31.1.93	Area 9			76.01	0.21	5.35			0.26						5.93	87.76
	Area 10			75.10	0.18	3.71			0.22			0.39			5.28	84.91
	Area 11		1.15	75.25	0.18	3.74			0.22			0.40			5.30	86.25
	Area 12		1.25	21.67		12.59						10.01			2.04	47.63
	Area15		1.36	22.20		13.73			0.27			10.12	0.24	2.04	2.07	51.85
Sample No 88	Area 1		89.54			1.57										91.11
MINERAL	Area 4	37.10	29.18			0.74	0.12	2.69				0.55				70.38
31.1.93																

Table No 8 EPMA/EDS Analysis, wt%, mineral, Mina Golpejas, Spain

The series of furnace experiments was undertaken to see if the complex, mixed mineral collected from the Golpejas mine would produce a natural bronze. Graphite crucibles were used and the tin/copper mineral was also covered with a layer of (carbon) graphite. Temperatures in the region of 1000°C to 1200°C were used on this experiment.

(Sample No 74) This ore sample weighed 1.46 grams. The sample was placed in a graphite crucible and completely covered with graphite. The sample was put in the furnace at a temperature setting of 500°C. The sample was visibly checked every half hour as the temperature was increased, but no visible change was noted. At two hours and twenty minutes the temperature reached 1200°C and this temperature was held till three hours and thirty minutes.

The sample was then removed from the furnace. The sample weighed 0.90 grams when removed from the furnace.

Metallic prills were visible including a large one (Sample No 74) when it was later examined with a optical microscope. A single area inside this prill was analysed on the SEM and it proved to be a natural bronze 6.25% Sn and 90.27% Cu.

(Sample No 82) This ore sample weighed 1.00 g. The sample was placed in a graphite crucible and covered with graphite. The Eurotherm furnace was pre-heated to 1100°C. This temperature was held for three hours The sample was then put into the furnace. The sample was kept in the furnace for one hour and twenty minutes and then removed and visibly examined. Metallic Prills were visible in some areas of the complex mineral.

(Sample No 82) Continued The furnace temperature was then increased to 1250°C. At one hour and thirty minutes the sample was put back into the furnace. The sample was finally removed from the furnace after 2 hours and ten minutes. The weight of the sample was 0.80 grams. Four small metallic prills areas were analysed in the SEM. All four areas contained a mixture of Sn (35%) and Cu (66%) with traces of Si, Fe and S also being present.

(Sample No 83) This ore sample weighed 0.47 grams. The sample was placed in a graphite crucible and covered with graphite. The Eurotherm furnace was used. The sample was put in the furnace at at a temperature of 1100°C. The furnace procedure time and temperatures used on this sample was exactly the same as for (Sample No 82). The sample weighed 0.53 grams when taken out of the furnace. Small metallic prills were visible on some surfaces of the mineral sample.

This sample was not analysed on the SEM as I felt it would only mirror the SEM results of (Sample No 82).

(Sample No 83) This ore sample weighed 1.63 grams when put in the furnace. The furnace procedure used on this sample was again a exact duplicate of (Sample No 82 & No 83). The sample was in the furnace for 1 1/2 hours at a temperature of 1100°C and the in the furnace for a further 1/2 hour at a temperature of 1250°C. I did not weigh the sample after removing it from the furnace. Small visible prills had developed on various areas of the sample.

The Celsius melting temperatures of the major elements contained in the complex tin/copper mineral:

Tin	(Sn)	0231.96 (°C)
Copper	(Cu)	1083.45 (°C)

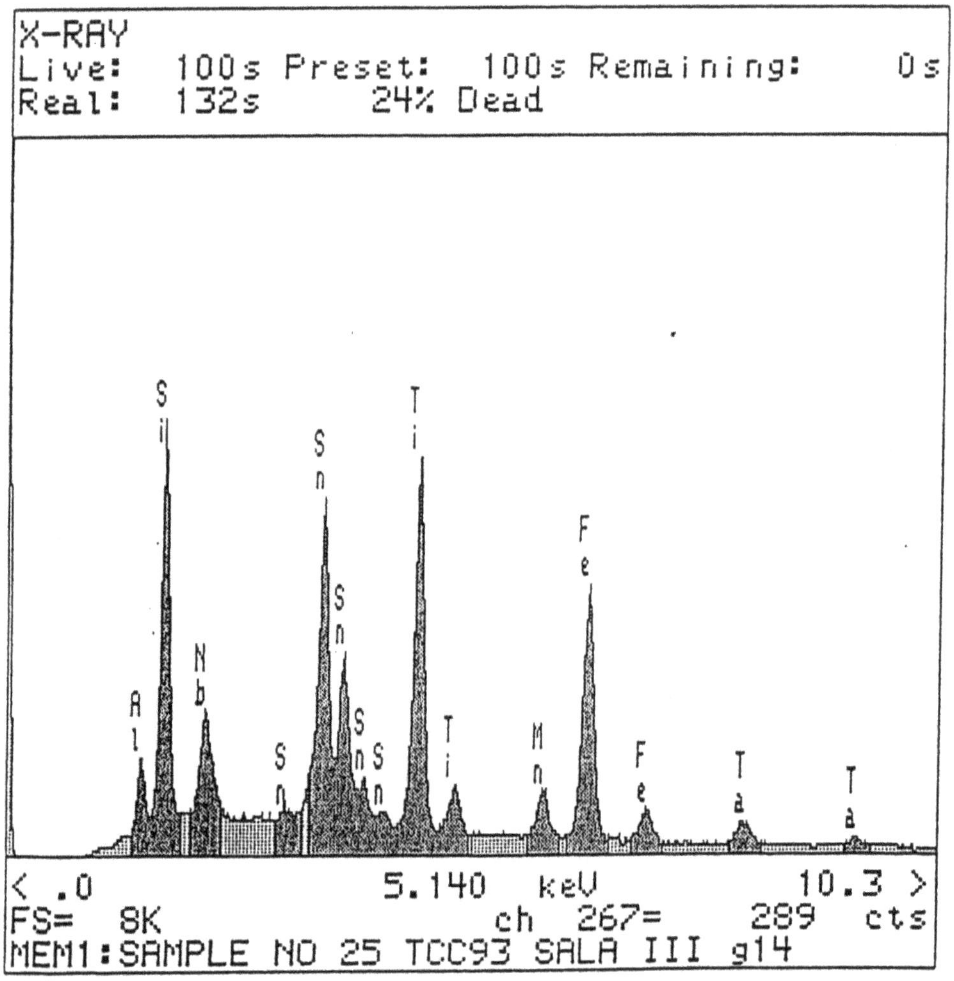

Graph No 1 SEM/EDS analysis slag Torre Romanan Centumcellas, Portugal

TORRE ROMANA CENTUMCELLAS

Visit No 20 (Pages 92-94)
SEM, and EPMA were used to analysis the Torre Romana Centumcellas slag samples. Ten samples of surface and excavated slag were analysed on the SEM. Seven of the analysed slags proved to be tin slag (Roman) and the other three pieces of slag were interpreted as iron slag also (Roman). All the tin slag samples contained numerous small tin prills.

Thirty general areas were analysed in the seven tin slag samples. The general tin content of each of the thirty analysed areas measured varied considerably from (2.2% Sn - 20% Sn). Most of the areas measured in the tin slags also contained varying percentages of the elements Niobium (Nb), Titanium (Ti) and Tantalum (Ta).

Photo 43 Ten pieces of surface Roman Tin slag, Torre Romana Centumcellas

TORRE		%	%	%	%	%	%	%	%	%	%	%	%	%	%	%
ROMANA		Sn	Nb	Ti	Ta	Al	Si	Fe	Mn	S	K	P	Ca	Mg	Na	TOTAL
CENTUMCELLAS																
Visit 20																
Sample No 1	Area 1	11.0	5.76	12.5	5.30	1.80	9.29	13.5	2.16							61.23
SLAG	Area 2	10.8	5.12	12.3	5.72	1.76	9.28	13.6	2.31							60.97
25.2.93	Area 3	11.0	5.59	11.8	5.87	1.92	9.88	13.6	2.29							62.09
Sample No 2	Area 1	14.8	5.73	10.31	5.65	2.39	10.9	13.0	2.11							64.91
SLAG	Area 2	17.1	6.24	9.38	6.16	2.56	11.7	11.7	2.07							66.88
25.2.93																
Sample No 7	Area 1	5.85	6.22	9.64	5.16	2.39	12.3	13.2	2.23							57.03
SLAG	Area 2	5.95	5.13	7.45	3.5	2.77	13	10.5	1.92							50.17
25.2.93	Area 3	7.1	2.74	4.26	1.32	3.7	14.5	10.8	2.23							46.70
	Area 4	6.73	4.05	7.75	2.7	1.92	5.51	16.4	2.18							48.09
	Area 5		14.3	27.5	16.7		9.12	3.98	0.3							71.92
	Area 6	2.24	12.2	18.9	12.7	0.87	12.3	11.2	1.41							71.87
Sample No 21	Area 1					0.16	0.19	90.5		0.06						90.98
SLAG	Area 2					2.60	11.0	43.4		0.08	1.06	0.27	0.30			58.84
11.5.93	Area 3					3.05	10.3	30.2		0.11	1.22	0.54	0.36			45.83
	Area 4					1.45	4.38	58.5		0.09	0.54	0.15	0.20			58.46
Sample No 22	Area 1			0.28		9.48	24.7	6.99	0.14	0.07	5.21		0.53	1.13	1.05	49.88
SLAG	Area 2	8.09	3.14	6.81	2.79	6.19	16.5	12.5	1.42	0.11	1.79		0.30	0.75	0.70	61.19
11.5.93	Area 3			5.11		4.37	21.3	15.9		0.08	2.53		0.39	1.28	0.53	51.62
	Area 4	2.13		2.27	0.63	8.73	21.2	11.6	1.15	0.09	3.42		0.27	0.61	0.56	53.17
Sample No 23	Area 1			0.31		5.26	15.7	34.9		0.12	1.4				0.49	58.3
SLAG	Area 2			0.54		8.70	18.8	21.3		0.10	2.23		0.09		1.85	54.59
11.5.93	Area 3			0.58		12.4	24	4.49		0.11	3.47				0.42	45.58
	Area 4			0.68		4.93	3.06	53.6			0.34					62.91
Sample No 24	Area 1					1.08	4.27	57.3		0.15	0.08		0.07			62.95
SLAG	Area 2					1.27	5.51	49.0		0.10	0.47					56.98
22.11.93	Area 3					1.62	9.63	37.6		0.16	0.59		0.07			50.89
	Area 4						0.21	50.4		0.19		0.95	0.18			51.13
	Area 5					5.00	11.4	36.2		0.07	0.61		0.13			53.56
Sample No 25	Area 1	18.8	5.72	11.4	6.20	1.79	9.45	12.9	2.12							68.37
SLAG	Area 2	21.6	5.75	9.96	5.69	2.18	10.29	12.07	2.00							69.58
16.12.93	Area 3	40.3	1.98	0.51	0.90	4.15	13.71	5.10	1.18							67.83
	Area 4	1.65	19.4	17.6	22.3		9.94	7.90	0.50							79.36
	Area 5	106.0				0.20	0.20									107.2
	Area 6	63.7				0.23	0.17									64.45
	Area 7	1.20	11.5	23.9	16.94		9.35	2.92	0.22							66.10
	Area 8	104.7		0.39		0.27	0.29	0.59								106.8

Table No 9 SEM/EDS Analysis, wt%, slag, Torre Romana Centumcellas, Portugal

TORRE		%	%	%	%	%	%	%	%	%	%	%	%	%	%	%
ROMANA		Sn	Nb	Ti	Ta	Al	Si	Fe	Mn	S	K	P	Ca	Mg	Na	TOTAL
CENTUMCELLAS																
Visit 20																
Sample No 26	Area 1	9.67	7.04	12.15	6.94	2.50	12.0	14.4	2.49							67.16
SLAG	Area 2	12.4	6.53	6.24	4.72	4.25	16.0	13.6	2.65							66.80
28.9.93																
Sample No 27	Area 1	7.25	6.15	10.0	5.59	2.06	10.7	12.1	2.30							56.23
SLAG	Area 2	6.77	6.52	10.8	7.00	2.05	11.0	12.2	2.33							58.66
16.12.93	Area 3		13.30	26.2	17.8		7.71	4.12	0.27							69.43
	Area 4	13.5	1.97	1.16	1.02	4.46	16.0	10.4	2.66							51.17
	Area 5	102.6				0.32	0.31	0.53								104.2

Table No 9 SEM/EDS Analysis, wt%, slag, Torre Romana Centumcellas, Portugal

(Sample No 1) Three areas were analysed on this homogeneous surface sample. The variation in the three areas analysed was negligible, all elements present Sn, Nb, Ti, Ta, Al, Si, Fe, and Mn did not vary more than one percent. Metallic tin prills were present in the slag.

(Sample No 2) Two areas were analysed on this surface slag sample and the analyses is almost a exact duplicate of (Sample No 1) in the elements and percentages present.

(Sample No 7) Six areas were analysed on this surface slag; four areas had a average of six per cent tin present.

Area 1 to Area 4 were similar to (Samples No 2 and No 27).

Area 5 was different in that it had no tin or aluminium present and high percentages of Nb (14.3%), Ti (27.5%), and Ta (16.7%) present and low percentage's of Fe (3.9%) and Mn (0.03%) present.

Point 6 had a small percentage of Sn (2.2%) and Al (0.8%) and high percentages of Nb (12.2%), Ti (18.8%) and Ta (12.9%) present.

(Sample No 21) Four points were analysed on this surface sample and the analytical results showed this slag as a product of an iron producing process.

(Sample No 22) Four areas were analysed on this very heterogeneous surface slag sample.

Area 1 analysis could represent an iron slag with a trace of Ti (0.3%) present.

Area 2 analysis represents a tin slag as in (Samples No 1, No 2 and No 7).

Area 3 represents an iron slag but with a significant percentage of Ti (5.1%) present.

Area 4 analysis represents a tin slag but with no Nb present and small amounts of Ti (2.3%) and Ta (0.6%) present.

(Sample No 23) Four areas were analysed on this surface slag sample and the analytical results as in (Sample No 21) showed this slag to be a product of a iron producing process, but with a minor variation trace amounts of Ti were present in all four areas.

(Sample No 24) Five areas were analysed on this piece of excavated slag. The analytical results match almost exactly with Sample No 21.

(Sample No 25) Eight areas were analysed on this piece of excavated slag.

Area 1 and Area 2 analysis can be what I would consider at present to be the standard analysis for the Roman tin slags.

Area 3 and Area 6 are also tin prills, but containing accompanying elements.

Area 4 and Area 7 analysis as a tin slag but with a small percentage of Sn present and large percentages of, Ti and Ta.

Area 5 and Area 8 are analyses of very pure tin prills embedded in the slag samples.

(Sample No 26) Two areas were analysed on this piece of excavated slag and both results confirmed what can be considered a typical analysis of the Torre Romana Centumcellas tin slags. The analysis had a good percentage of Sn 10% or higher and percentages of Nb, Ti, Ta, in the region of 5% or higher with varying percentages for the accompanying elements Al, Si, Fe and Mn.

(Sample No 27) Five areas were analysed on this excavated slag sample.

Areas 1 and 2 represent the standard analysis for this Roman tin slag.

Areas 3 analysis has no tin present but contains high percentages of Nb (13.3%), Ti (26.2%) and Ta (17.8%) with the accompanying elements Si, Fe and Mn.

Area 4 also represents the standard analysis but with a high percentage of Sn (13.5%) and Fe (10.4%) and small percantages of the other accompanying elements.

Area 5 is tin prill with small trace elements of Al, Si, and Fe present.

The Torre Romana Centumcellas slag was also analysed on the Electron Microprobe at the Institute of Archaeology to check and confirm the results of the SEM analysis. (Sample No 26) was the sample analysed on the Microprobe.

(Sample No 26) Numerical percentages were not acquired: only graphs were produced; showing the presence and intensity of the characteristic X rays of the elements.
A series of photographs showing the different phase compositions and the areas analysed was also produced.

File No 25 [1st scan], 5 micron area, Ti, Ta, Fe, and Al.

File No 26 [2nd scan], 5 micron area, Mn, Sn, K, Nb, Fe, Si,and Al.

File no 27 [3rd scan], 5 micron area, Mn, Ti, Ta, Fe, Al, and Mg.

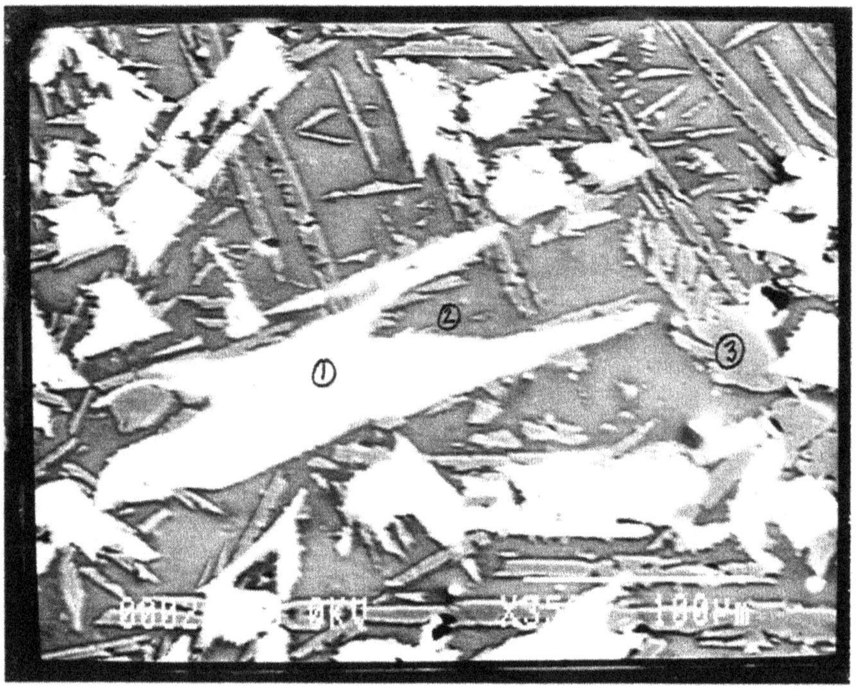

Photo 44 Microprobe photograph Roman Tin Slag, Torre Romana Centumcellas

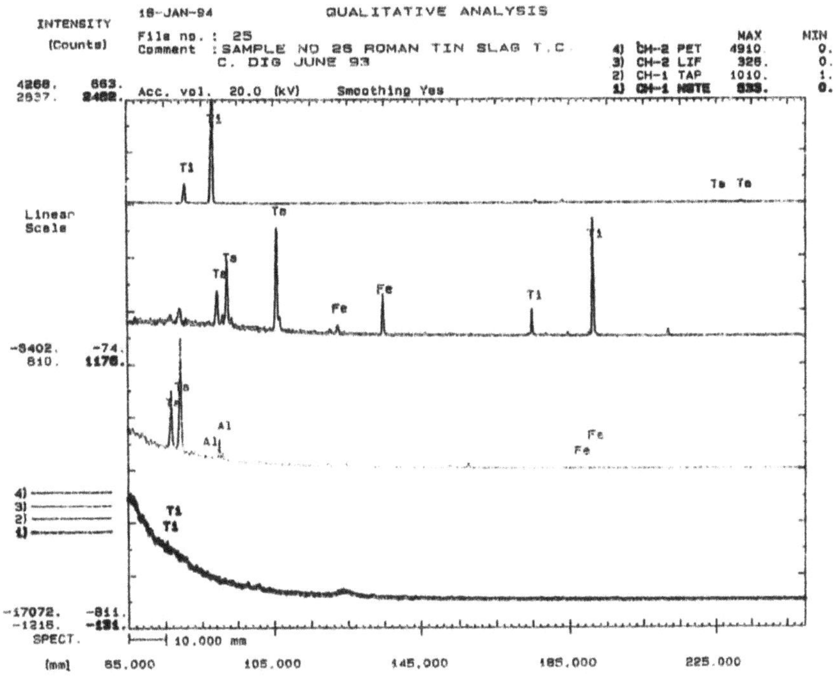

Graph No 2 Microprobe analysis slag, Microprobe file no 25, area 1, Torre Romana Centumcellas, Portugal

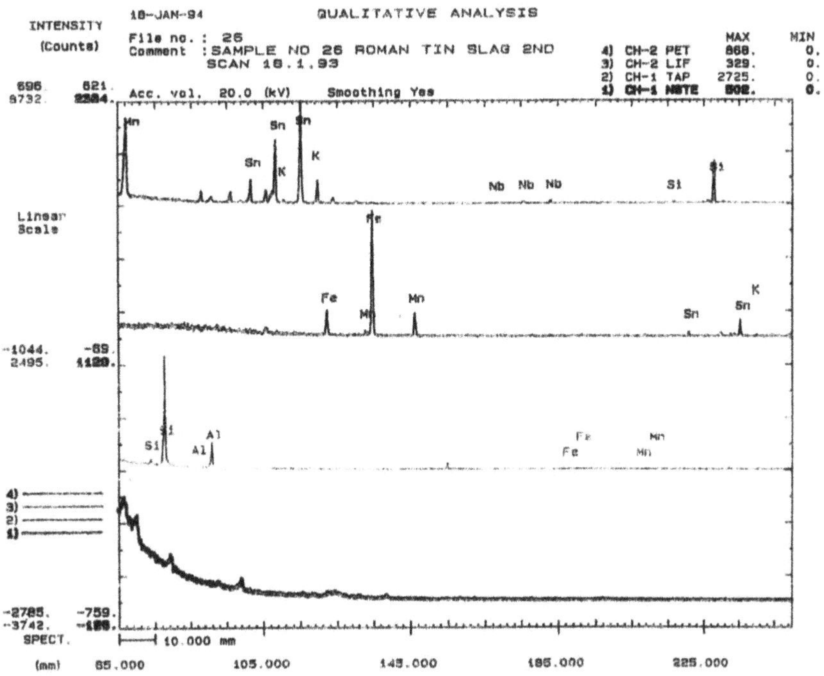

Graph No 3 Microprobe analysis slag, Microprobe file no 26, area 2 Torre Romana Centumcellas, Portugal

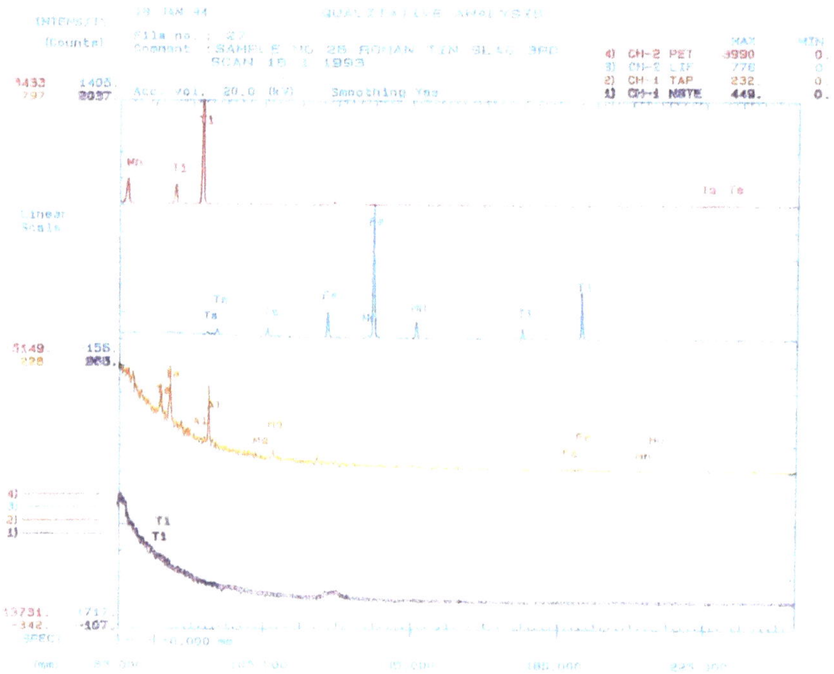

Graph No 4 Microprobe analysis slag, Microprobe file no 27, area 3 Torre Romana Centumcellas, Portugal

The Microprobe at the Institute of Archaeology was also used to scan and map the elements (Sn, Nb, Ti, Ta, Al, Si, Fe, and Mn) present in (Sample No 26). At a later date varied elements and element combinations of the stored results of the microprobe mapping/scan were brought up on the microprobe coloured monitor screen and photographed.

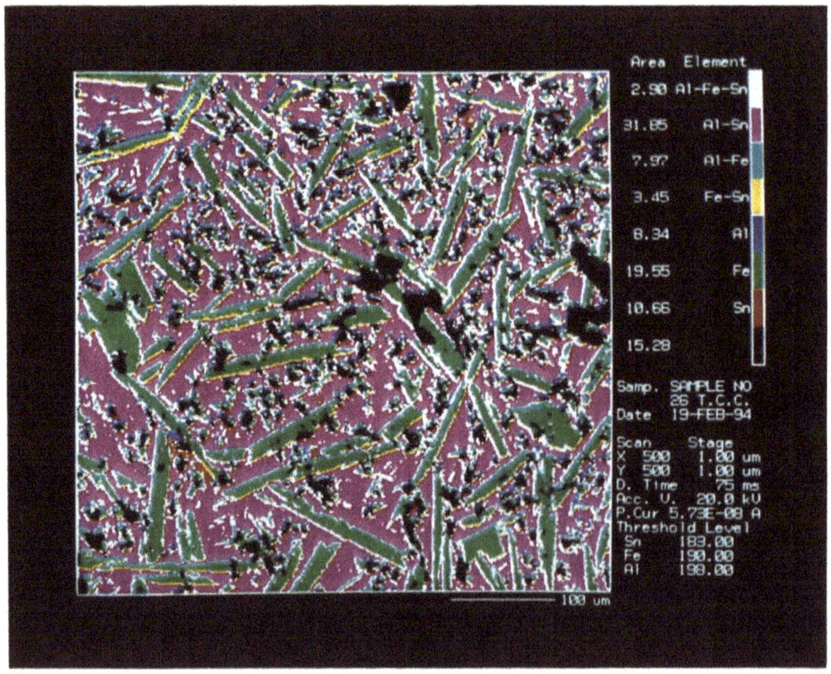

Photo 45 Coloured Microprobe Configuration Scan, Roman Tin Slag, Sample No 26, Torre Romana Centumcellas

A series of furnace experiments was done using the Roman Tin slag; and was undertaken to try to establish the melting point of the combination of elements Sn, Nb, Ti, Ta, Fe, Al, Si and Mn which make up the Roman Tin Slag. The Institute of Archaeology laboratory furnaces were used for all the experiments. The sample numbers represent my PhD laboratory work sequence of the samples.

All the seven small pieces of furnace experiment slag were cut or broken from bigger samples of the analysed Torre Romana tin slag. All the cut samples were put in small graphite crucibles and completely covered with a layer of graphite. The crucibles were then placed in either a Anderman or Eurotherm ceramic furnace at various temperatures and for various periods of time.

The temperatures used in the slag experiment were what is thought were the general temperatures that metallurgists could have been capable of achieving in antiquity. The usual antiquity smelting and smithing temperatures would have been in the range of 1000°C to 1200°C (Tylecote, 1992, 48-9). Hans-Gert Bachmann states, "The temperatures in early smelting furnaces and hearths rarely exceeded 1300°C-1400°C; in most cases they were lower" (Bachman, 1982, 10).

(Sample No 75) The weight of the slag sample was 1.30 grams when the sample was put into the furnace which had been preheated to 500°C. The furnace temperature was increased slowly by 100°C intervals until it reached 1200°C at 2 hours 20 minutes. The crucible sample was checked every half hour to see if plasticity or melting had taken place. The temperature of 1200°C was held till 3 hours 30 minutes, after the start, at this time no plasticity or melting of the slag sample was visible. When the sample was taken from the furnace it weighed 1.21 grams.

(Sample No 76) The weight of the slag sample was 2.07 grams when the sample was put into the furnace. The primary experimental procedure was the same as (Sample No 75). At 2 hours 10 minutes hours the furnace temperature reached 1100°C. At this time there was no visible change in the sample. The furnace reached a temperature of 1200°C at 2 hours 20 minutes. This temperature was held till 3 hours 10 minutes had passed. The sample by this time had started to become visibly plastic and soft, but was still not fully molten. The sample weighed 1.99 grams when taken from the furnace at 3 hours and 10 minutes.

(Sample No 78) The weight of this sample was 0.44 grams. The sample was put into the furnace at the temperature of 1335°C. The sample was removed from the furnace at 24 minutes. The weight of the sample was 0.40 grams. No physical change of the sample had taken place due to the short time the sample was left in the in the furnace, 23 minutes.

(Sample No 79) The weight of this sample was 0.39 grams. The sample was put into the furnace when it had reached a temperature of 1420°C. The sample was left in the furnace for only (32 minutes). The sample weighed 0.34 grams when removed. Visually the sample had changed considerably, it had gone very plastic and had started melting before it was removed from the furnace. Given longer time in the furnace at this high temperature there can be no doubt that the Roman Tin slag would have completely melted.

(Sample No 85) This Sample weighed 0.44 grams when put into the furnace. The Eurotherm furnace had been running for three hours at a constant temperature of 1100°C when the sample was put into the furnace. After one hour the furnace was set to 1250°C and at one hour and thirty five minutes the temperature was again raised to a setting of 1285 °C. The sample was removed from the furnace after two hours. No visual change had taken place, the sample had not turned plastic and still retained its sharp, jagged edges.

The Celsius melting temperatures of the elements contained in the Roman Tin slag are as follows:

Aluminium	(Al)	660.37°C
Iron	(Fe)	1534.85°C
Manganese	(Mn)	1243.85°C
Niobium	(Nb)	2467.85°C
Tantalum	(Ta)	2995.85°C
Tin	(Sn)	231.96°C
Titanium	(Ti)	1659.85°C
Silicon	(Si)	1409.85°C

MINA DE SANTA EULIA - Visit No 25 (Pages 100-102)
Two samples were analysed from this site: one mineral and one slag.

(Sample No 10) This mineral sample analysed as a very rich piece of cassiterite (75% Sn).

(Sample No 18/B) This slag sample was analysed on the SEM in the hope of finding traces of metallic tin within the slag. The analyses were negative and it seems to be a iron slag (average 20% Fe). This sample is possibly the remains of a iron forging/smithing process; possibly Roman. As possible Roman Tegula roof tiles were found in the same area the slag.

MINA DE SANTA EULALIA		%	%	%	%	%	%	%	%	%	%	%	%	%	%	%	%
		Sn	Cu	Nb	Ti	Ta	Al	Si	Fe	Mn	S	K	P	Ca	Cl	Na	TOTAL
Visit 25																	
Sample No 10	Area1	77.07															77.07
MINERAL	Area 2	74.74															74.74
5.2.93																	
Sample No18/B	Area 1						8.71	23.74	1.75		0.09	1.02		2.97		2.73	41.16
SLAG	Area 2				0.21		5.34	14.06	18.16		0.15	1.54		0.53	0.18	1.09	41.57
10.2.93	Area 3							38.90	0.32		0.11					0.47	40.22
	Area 4				0.08		4.75	13.34	42.81			2.79		0.74		1.09	65.86
	Area 5						0.23	0.95	51.05		0.12				0.10		52.94

Table No 10 SEM/EDS analysis, wt%, slag, mineral, Mina de Santa Eulalia, Portugal

MINA VAL DE FLORES		%	%	%	%	%	%	%	%	%	%	%	%	%	%	%	%
		Sn	Cu	Nb	Ti	Ta	Al	Si	Fe	Mn	S	K	P	Ca	Mg	Na	TOTAL
Visit 29																	
Sample No 40	Area 1						3.61	4.01	40.22			0.51	0.22				48.59
MINERAL(?)	Area 2						2.71	4.61	38.31			0.64	0.15				46.44
2.12.93																	
Sample No 41	Area 1		2.09		0.19		12.64						14.67	3.48		0.41	33.59
MINERAL (?)	Area 2		1.69		0.11		11.23	0.11					13.03	2.86		0.72	29.81
2.12.93																	
Sample No 42	Area 1	69.17							0.32								69.50
MINERAL	Area 2	70.38							0.19								70.57
1.12.93	Area 3	68.06							0.11								68.17
Sample No 43	Area 1						1.52	1.83	54.51			0.10	0.28				58.26
MINERAL	Area 2						11.85	11.79	30.92								54.58
2.12.93	Area 3						0.31	0.93	59.57								60.91
Sample No 39	Area 1				0.45		7.93	23.84	9.71	1.00	0.14	2.19		3.18	2.38		50.85
SLAG	Area 2				0.76		7.78	16.41	6.65	0.48	0.14	0.13		10.36	7.36		47.71
29.9.93																	
Sample No 71	Area 1				0.41		8.04	21.34	7.84	1.00		0.82		4.47	3.80		47.76
SLAG	Area 2				0.46		7.55	21.00	10.35	0.74		2.27		3.28	2.41		48.09
20.1.94	Area 3				0.50		8.30	22.69	7.94	0.79		0.49		9.84	5.99		56.57
	Area 4				0.44		8.58	24.26	8.65	0.91		1.36		4.42	3.29		51.94
	Area 5				0.26		4.48	24.13	5.35	0.49		2.05		0.67	0.56		38.02

Table No 11 SEM/EDS analysis wt%, mineral, slag, Mina Val de Flores, Spain

MINA VALDEFLORES

Visit No 29 (Pages 105-108)
Six samples were analysed from the Mina Val de Flores; four of these samples were possible mineral samples and two sub samples were cut from a single small piece of slag.

(Sample No 40) The analytical results proved this sample was possibly iron ore.

(Sample No 41) This sample had a small percentage of copper present (1.8% Cu) and traces of titanium (0.1% Ti). The analysis of this sample shows that small amounts of copper are present at Val de Flores.

(Sample No 42) This sample is a rich cassiterite ore (69% Sn).

(Sample No 43) This sample is a iron ore mineral.

(Sample No 39) This sample analysed on the SEM is a slag, but most probably from iron smelting. The iron content is low (8.0% Fe) with a small trace of Titanium (0.5%) also present. I had hoped to find some tin in this slag which would have offered proof that tin smelting had been taking place at Val De Flores.

(Sample No 71) This sample is from the same slag as (Sample No 39). I analysed five areas on this sample in the hope of locating some small trace of tin within the slag structure, but the results were negative. The analysis of this sample is interpeted as a iron slag, but with a low iron content averaging around (8.0% Fe).

CATRAIA DO		%	%	%	%	%	%	%	%	%	%	%	%	%	%	%	%
BURACO		Sn	W	Nb	Ti	Ta	Al	Si	Fe	Mn	S	K	P	Ca	Mg	Na	TOTAL
Visit 30																	
Sample No 28	Area 1						1.62	10.07	51.25	0.42	0.08	0.57	0.28	0.23		0.56	65.12
SLAG	Area 2						3.43	9.28	53.8	0.52	0.10	1.05	0.33	0.49		0.75	69.79
27.1.94	Area 3						6.40	13.33	38.23	0.23	0.22	4.17	1.74	0.88		1.57	66.81
	Area 4						6.25	13.34	41.2	0.49	0.16	1.86	1.10	1.53		2.56	68.52
	Area 5						5.23	11.16	33.68	0.36	0.12	1.86	0.96	1.19		2.11	56.99
	Area 6						3.17	11.19	39.12	0.34	0.07	0.86	0.53	0.83		1.21	57.49
	Area 7						0.52	11.33	44.04	0.61		0.51		0.19		0.68	58.01
	Area 8						1.02	1.71	58.13	0.18		0.52		0.33		0.50	63.04
Sample No 29	Area 1						2.87	8.62	47.75	0.45	0.10	0.85	0.35	0.35			61.36
SLAG	Area 2						1.70	8.85	49.34	0.52	0.09	0.63	0.18	0.30			61.63
23.9.93	Area 3						2.83	6.45	41.23	0.38	0.17	0.92	0.19	0.40			52.62
Sample No 30	Area 1						4.37	17.36	34.38	0.19	0.13	2.06		0.83		0.87	60.29
SLAG	Area 2						3.66	12.95	46.97	0.21	0.06	1.79		0.90		0.97	67.66
27.10.93	Area 3						1.96	4.27	45.95		0.21	0.44		0.08		0.23	53.54
	Area 4						4.01	11.93	49.39	0.21		1.67		0.51		1.00	68.86
	Area 5						3.29	11.43	50.99	0.23	0.07	1.52		0.50		1.18	69.33
Sample No 99	Area 1						8.19	13.2	22.02			5.45	0.35			0.59	49.81
SLAG	Area 2						1.72	7.42	51.58			0.13	0.43			0.40	61.70
15.11.94	Area 3						1.38	6.12	53.58			0.46	0.20			0.43	62.18
	Area 4						0.60	0.14	66.28								67.77
	Area 5						0.70	0.33	28.44				0.37				29.95

Table No 12 SEM/EDS analysis, wt%, slag, Catraia do Buraco, Portugal

CATRAIA DO BURACO

Visit No 30 (Pages 109-111)
This site produced pottery, crushing stones, hammer stones,and slags. At present a firm archaeological date has not been given to the site due to the fragmentary nature and the possible multiperiod use that the surface finds represent.

Four slag samples were analysed from this site:

The four slag samples were analysed on the SEM/EDS (Sample No 28, Sample No 29, Sample No 30 and Sample No 99) in the hope of finding traces of tin within the slags. A total of twenty one areas were analysed on all four slags, all the analysis pointed to iron production slags (average iron content 45%-50%). No traces of tin, wolframitte, titanium or tantalum were found in any of the twenty-one areas analysed.

The Catraia Do Buraco site is located only a few hundred meters to the North-East of Torre Romana Centumcellas a site which produced Roman tin slags containing percentages of Sn, Ti, Ta, and Nb.

Most of the mineralzed tin deposits in the locality of Catraia Do Buraco are known to contain at least one of the three accompanying elements (Nb, Ti, Ta) and in a majority of the tin deposits all three of the elements (Carta Geologica De Portugal, SABUGAL 21-A, 1:50.000, 1964).

It could be expected that the Catraia Do Buraco smelting slags would contain at least some trace of tin and one or more of the elements (Nb, Ti, Ta) regrettably none appear. The iron slag being produced on what looks like a possible Late Bronze Age site poses a problem.

This problem could be resolved by a number of arguments. Iron production may in certain locations in Mid-Iberia started earlier than previously thought or the period of transition from Bronze to Iron may have lasted longer than prevously thought.

Until a firm date for the Catraia Do Buraco site is obtained the above arguments are meaningless. Obviously in the future a small excavation for obtaining stratigraphy needs to be done. The site is being threatened due to a building extension of a nearby factory.

BELMONTE		%	%	%	%	%	%	%	%	%	%	%	%	%	%	%	%
CASTLE		Sn	Nb	W	Ti	Ta	Al	Si	Fe	Mn	S	K	P	Ca	Mg	Na	TOTAL
Visit 36																	
Sample No 31	Area1							11.98	42.57		0.09	2.37		0.46	0.28		57.76
SLAG	Area 2							11.26	42.49		0.05	1.70		0.71			56.24
28.9.95																	
Sample No 32	Area 1						3.88	10.13	35.59		0.07	1.69		0.54		0.41	52,39
SLAG																	
20.1.94																	
Sample No 33	Area 1						3.08	4.72	60.42		0.16	0.63		1.06	0.35		70.44
SLAG	Area 2						5.34	11.89	47.61		0.26	1.32		2.18	0.53		69.16
22.11.93	Area 3						7.04	9.49	49.35		0.25	0.83		1.53	0.79		69.32
Sample No 34	Area 1				0.20		5.08	13.47	29.94	0.46		3.89	0.33	1.99	0.38	1.02	56.8
SLAG																	
20.1.94																	
Sample No 35	Area 1						3.47	10.38	42.8	0.79	0.15	1.22	0.17	0.35			59.35
SLAG																	
20.1.94																	
Sample No 36	Area 1						4.61	14.83	32.68			2.50		0.79			55.56
SLAG																	
20.1.94																	
Sample No 53	Area 1						2.02	9.28	47.79	4.23		0.62	0.45				64.42
SLAG	Area 2						2.11	9.60	49.81	4.39		0.69	0.50				67.12
20.1.94	Area 3						2.13	9.78	47.31	4.51		0.63	0.48				64.86

Table No 13 SEM/EDS analysis, wt%, slag, Belmonte Castle, Portugal

BELMONTE		%	%	%	%		%	%	%	%	%	%	%	%	%	%	%
CASTLE		Sn	Nb	W	Ti	Ta	Al	Si	Fe	Mn	S	K	P	Ca	Mg	Na	TOTAL
Visit 36																	
Sample No 54	Area 1						3.29	9.07	38.53	0.54		1.50		0.37			53.42
SLAG	Area 2																
13.1.94																	
Sample No 55	Area 1						3.57	10.08	41.59			1.22	0.23	0.53		0.69	57.94
SLAG																	
13.1.94																	
Sample No 56	Area 1						2.39	8.08	44.0	1.20	0.13	0.86	0.60	0.34			57.63
SLAG																	
13.1.94																	
Sample No 57	Area 1						0.68	1.26	67.23								69.18
SLAG	Area 2						0.79	1.86	58.86								61.51
13.1.94																	
Sample No 58	Area 1				0.29		4.39	12.14	41.75			1.60		0.34			60.65
SLAG																	
13.1.94																	
Sample No 59	Area 1						1.49	9.68	47.51	2.50		0.37	0.34				61.91
SLAG																	
13.1.94																	
Sample No 60	Area 1						4.04	12.41	33.71	0.17			1.62	0.55			52.52
SLAG																	
13.1.94																	
Sample No 61	Area 1				0.12		5.93	13.6	31.93			2.30		0.41	0.24		54.55
SLAG	Area 2				0.08		3.85	11.97	36.72			1.03		0.37	0.28		54.32
13.1.94																	
Sample No 62	Area 1						3.50	10.04	41.98		0.12	0.54		0.30			56.57
SLAG																	
20.1.94																	
Sample No 63	Area 1						5.19	11.49	29.97	0.35		2.73	0.94	5.47	0.25		56.82
SLAG					0.33												
13.1.94																	
Sample No 64	Area 1						4.75	10.99	47.19	0.63		2.96	1.09	0.61			68.25
SLAG	Area 2						3.33	11.08	48.82	0.81		2.06	1.03		0.82		67.98
22.11.93	Area 3						7.67	11.66	38.42	0.56		4.30	1.74	1.29			65.67
Sample No 65	Area 1						2.60	8.92	34.64	0.45	0.17	1.21	0.20	0.36			48.57
SLAG																	
13.1.94																	
Sample No 66	Area 1						2.75	7.64	36.13	0.74		0.84	0.43	0.29			48.85
SLAG																	
13.1.94																	

Table No 13 SEM/EDS analysis, wt%, slag, Belmonte Castle, Portugal

BELMONTE		%	%	%	%	%	%	%	%	%	%	%	%	%	%	%	%
CASTLE		Sn	Nb	W	Ti	Ta	Al	Si	Fe	Mn	S	K	P	Ca	Mg	Na	TOTAL
Visit 36																	
Sample No 67	Area 1						3.70	8.42	40.97	0.28	0.32	0.98		0.39			55.3
SLAG					0.20												
20.1.94																	
Sample No 68	Area 1						3.50	10.2	46.68	1.09		2.14		0.34	0.29		64.37
SLAG					0.10												
20.1.94																	
Sample No 69	Area 1						5.47	14.02	36.91	0.16		2.35		1.28		0.65	60.95
SLAG					0.08												
20.1.94																	
Sample No 70	Area 1						3.98	11.43	36.08	0.21		2.67	1.08	6.91	0.33		62.94
SLAG					0.23												
20.1.94																	

Table No 14 SEM/EDS analysis, wt%, slag, Belmonte Castle, Portugal

BELMONTE CASTLE

Visit No 36 (Page 115)
Thirty-three areas on twenty-four separate slag samples were analysed on the SEM/EDS. The reason for collecting and sampling such a large collection of slag was the hope of finding some traces of Sn or the elements Ti, Ta, Nb.

Unfortunately all of the thirty three analysed areas of the Belmonte slag samples analysed as iron smelting slags with a average iron content of around 40%. Nine areas in eight of the samples had small traces of titanium, which could point to a local iron ore being used. No traces of tin appeared in any of the slag samples.

Photo 46 assortment of collected surface slag, Belmonte Castle

CRIFT FARM, CORNWALL

(Sample No 11) This sample a small piece of alluvial cassiterite from Cornwall, England which was collected and given to me by John Merkel, which was analysed to check the percentage of tin present. The alluvial cassiterite sample proved to be rich in tin (82% Sn).

CRIFT FARM		%	%	%	%	%	%	%	%	%	%	%	%	%	%	%	%
CORNWALL		Sn	Cu	Nb	Ti	Ta	Al	Si	Fe	Mn	S	K	P	Ca	Mg	Na	TOTAL
Sample No 11	Area 1	82.70															82.70
MINERAL	Area 2	83.03															83.03
ALLUVIAL TIN																	
9.2.93																	

Table No 15 SEM/EDS analysis, wt%, mineral, Crift Farm, Cornwall, England

PYLMTON, DEVON

(Sample No 72) This sample of "wood tin" from Devon, England, which was purchased at a mineral fair in London was analysed to determine the percentage of tin present. The wood tin sample proved to be rich in tin (average 60% Sn).

PLYMTON		%	%	%	%	%	%	%	%	%	%	%	%	%	%	%	%
DEVON		Sn	Cu	Nb	Ti	Ta	Al	Si	Fe	Mn	S	K	P	Ca	Mg	Na	TOTAL
Sample No 72	Area 1	76.32						0.23	1.30								77.85
MINERAL	Area 2	50.93						23.02	0.48								74.44
WOOD TIN	Area 3	0.28						40.56									40.90
9.12.93																	

Table No 16 SEM/EDS analysis, wt%, mineral, Plymton, Devoon England

Excavated Late Bronze Age crucible fragments have been given to me by Raquel Vilaca of the Instituto de Arqueologia, Coimbra University, Portugal and she has given me her permission to include the crucible fragment analytical results in my PhD.

The crucible fragments come from two Late Bronze Age sites Castelejo and Alegrios, which Raquel Vilaca has been excavating seasionally over the last five years.

The Vilaca site of Castelejo is in the Mid-Eastern Portuguese province of Beira Baxia, Administrative District of Sabugal, at 1008 meters altitude, just above the small village of Quarta-Feira.

North: 40° 20' 06" - East of Lisboa: 01° 20' 54"
Carta de Portugal; Beiras Folha 4, 1:200.000
Instituto Geografico e Cadastral, 988

All the analysis work on the crucible fragments was done using (EPMA/EDS) analysis on the microprobe in the Geology Department, Birkbeck College. The microprobe technician was Andy Beard. A quantitative computer program was used to acquire the analytical results. Black and White photographs were also taken of all the analysed areas.

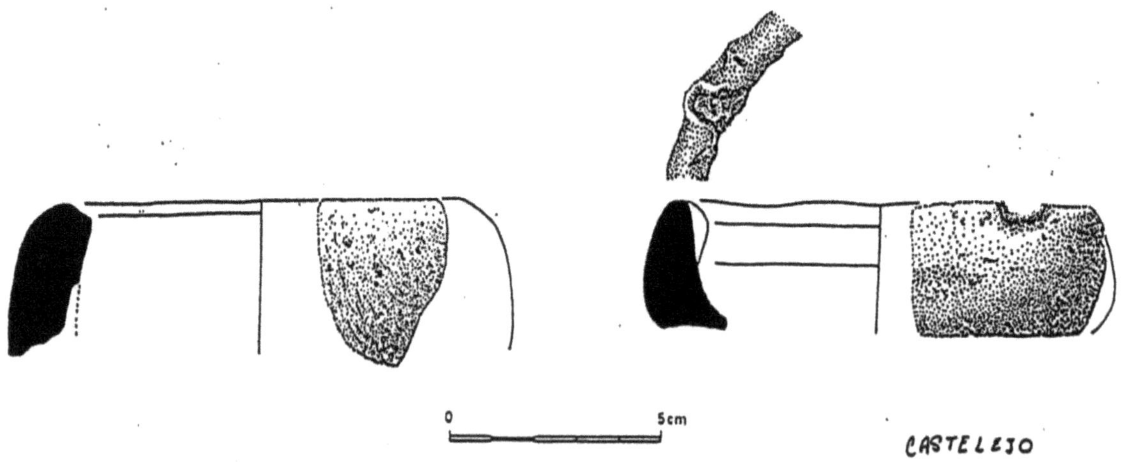

Drawing No 22 crucible fragments, Castelejo, Portugal, Raquel Vilaca, scale units 1 cm

The crucible samples were sectioned, mounted and polished. A total of seven separate samples were cut from the Castelejo crucibles. After visual examination of the seven mounted/polished samples using an optical microscope, two of the seven were chosen for microbrobe EDS analysis.

(Sample Cast No 10) Area 1 and Area 2 Two micro tetragonal shaped fragments located just below the rim of the inner wall of the crucibles were analysed. Both yielded almost identical analysis. The micro points analysed to almost pure tin (73.3% and 70.4%) with small percentages of copper present(1.3% and 0.8%).

The other elements present are Al (minor 1-5%), Si (minor 1-5%), Ca (minor 1-5%); Fe (trace, less than 1%),and P (trace, less than 1%) and . The trace elements probably come from the pottery or ceramic material used to make the crucible.

(Sample Cast No 11) Six areas were analysed on this crucible sample.

Area 1 A tetragonal shaped fragment located just below the rim of the crucible and below the inside surface (one of many similiar fragments), contained (80.2%) Sn. This point is a fragment of cassiterite ore. The missing percentage to 100% would consist of oxygen which the microprobe does not detect. It is not smelted tin. I would consider the fragment to be a natural cassiterite mineral that has been crushed and put into the crucible.

The crushed cassiterite fragment become embedded in inner area of the upper crucible wall, which seems to have become very porous and fragmented during the bronze making process.

Area 2 This area is a similiar to Area 1 expect for a slightly higher tin content (82.7%) Sn and a trace of (0.1%) Si.

Area 3 This area is also similiar to Area 1 Sn (79.9%) and traces of Al (0.2%) and Si (0.2%).

Area 4 This area is also similiar to Area 1 Sn (81.2%) and Si (0.1%).

Area 5 This area is a metallic prill. The location of the prill is just below the rim and on the inside wall of the crucible, in the same general area as the cassiterite fragments Area No 1 - Area No 4. The analytical results of this prill showed it to be 99% copper with a trace of Si (0.1%). This analysis shows that metallic copper was present in the crucible along with cassiterite ore.

Area 6 This area is identical to Area 5 and contains Cu (94%) with a trace of S (0.5%).

The above analytical results along with the Black and White photographs point to a very strong possibility that the Castelejo Crucible (Cast No 11) was used for a bronze alloying process.

The second excavated Vilaca site of Alegrios is in the Mid-Eastern Portuguese province of Beira Baixa, Administrative District of Idanha-a-Nova, located at approximately 600 meters altitude in the Serra dos Alegrios, approximately three kilometres to the north/west of the town of Monsanto. The second site is located approximately ten kilometers to the South of Castelejo.

North: 40° 00' 24" - East of Lisboa: 01° 59' 57"
Carta Portugal; Beiras Folha 4 1:200.000
Instituto Geografico e Cadastral; 1988

This site produced one crucible rim fragment, which was sectioned mounted and polished to make two samples from this single fragment.

RAQUEL VILACA		%	%	%	%	%	%	%	%	%	%	%	%	%	%	%	%
SAMPLES		Sn	Cu	Nb	Ti	Ta	Al	Si	Fe	Mn	S	K	P	Ca	Mg	Na	TOTAL
Cast No 10	Areal	73.29	1.28				1.46	3.55	0.42				0.24	1.00			81.26
CRUCIBLE	Area 2	70.39	0.79				2.07	4.46	0.31				0.21	1.22			79.43
FRAGMENT																	
14.9.94																	
Cast No 11	Area 1	80.23															80.23
CRUCIBLE	Area 2	82.75						0.12									82.75
FRAGMENT	Area 3	79.90					0.16	0.21									80.27
14.9.94	Area 4	81.20						0.17									80.37
	Area 5		100					0.14									100.1
	Area 6		94.0								0.48						94.55
Ale No 37/A	Area 1	29.90		1.13	0.37		4.73	15.7	8.07	1.77							62.00
CRUCIBLE																	
FRAGMENT																	
14.9.94																	
Ale No 37/B	Area 1	19.54	44.0						0.62		0.26						64.43
CRUCIBLE	Area 2	29.89		1.13			4.73	16.0	8.07	1.77							61.42
FRAGMENT	Area 3	0.32		9.45		28.7		2.49	6.49	1.03							48.70
14.9.94	Area 4	0.53		8.83	13.1	29.3	0.12	2.97	6.18	1.17							62.90

Table No 17 EPMA/EDS analysis, wt%, crucible fragments, Castelejo and Alegrios, Portugal

(Sample Ale 37/A) One area was analysed on this sample. The analysed area contained Sn (29.9%), Nb (1.1%), Ti (0.4%) and the accompanying elements Al (4.7%), Si (15.7%), Fe (8.1%) and Mn (1.8%).

The area analysed on Sample Ale 37/A was possibly too large and included not only a cassiterite fragment but an area of the crucible wall structure, which must account for the elements Al, Si, Fe, and Mn. What is interesting is the appearance of the elements Nb and Ti which appear geologically along with cassiterite in some areas of the Mid-Iberian region.

(Sample Ale 37/B) Five areas were analysed on this sample.

Area 1 The area analysed was a metallic prill located on the inner wall of the crucible. The analysis from Area 1 offered proof that bronze was being produced in the Alegrios crucible. Present was Sn (19.5%) and Cu (44.%) and traces of Fe (0.6%) and S (0.3%).

Area 2 This area is identical with the area analysed on (Sample Ale 37/A).

Photo 47 Crucible Fragment, Alegros excavated 1988, Raquel Vilaca

(Sample Ale 37/B)

Area 3 The area analysed at this point was a small irregular inclusion located on the lower inner wall of the crucible sample. The analysis showed a small trace of Sn (0.3%) present,with major percentages of Nb (9.4%), Ta (28.7%) and Fe (6.5%) present and minor percentages of Si (2.5) and Mn (1.0%). The analysis clearly shows that the fragment is a piece of Nb, Ti and Ta mineral with accompanying Fe, which has been mistakenly collected, crushed and put in the crucible along with the cassiterite ore.

Area 4 The area analysed in this point was another area within the same fragmented piece of mineral where Area 3 was analysed. The results were similar to Area 3 expect for the appearance of a major percentage of Ti (13.1%). This second analysis along with the black and white photographs leaves no doubt that local Iberian cassiterite ore was being added to copper in the Alegrios crucible to produce bronze for casting.

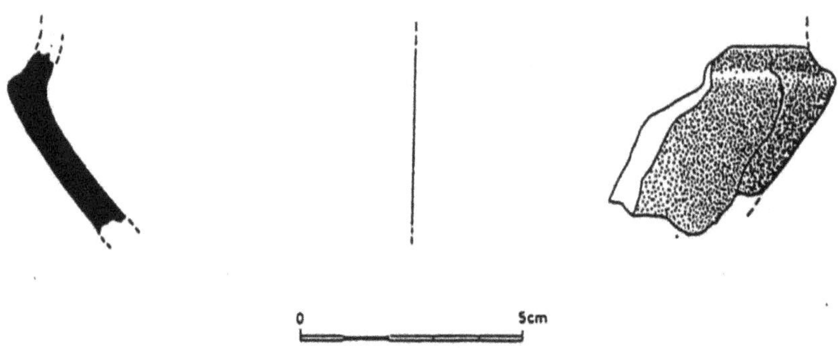

Drawing No 23 crucible fragment, Alegrios, Portugal, scale units 1 cm

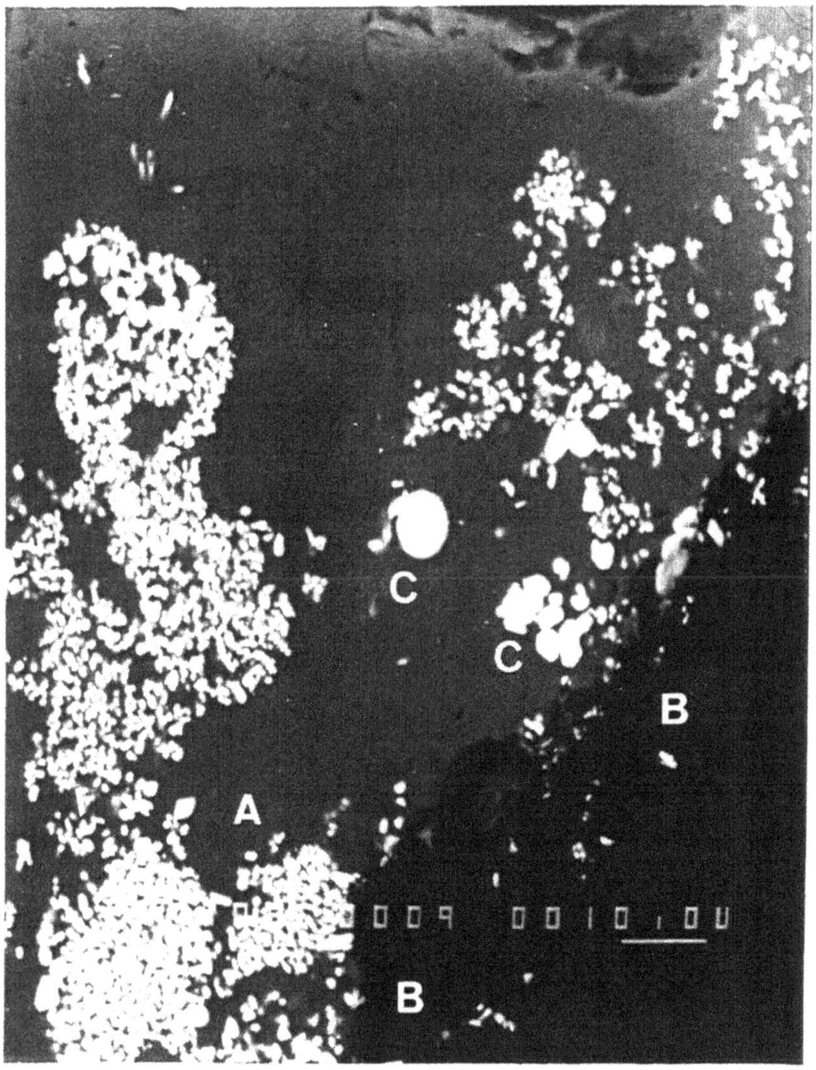

Photo 48 Microprobe photograph section of Raquel Vilaca crucible

(A) crushed cassiterite fragments, Cast No 11, areas 1-4, page 157

(B) inner upper wall of crucible

(C) copper prills, Cast No 11, Areas 5-6, page 157

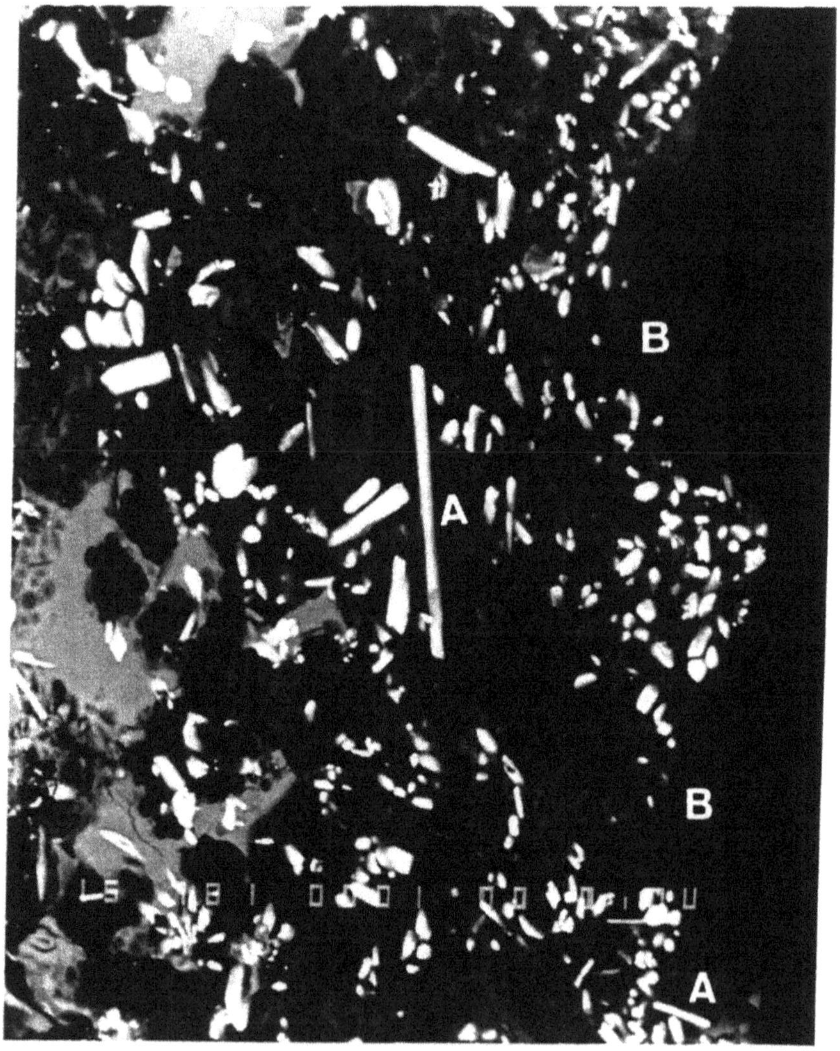

Photo 49 Microprobe photograph section of Raquel Vilaca crucible

(A) crushed cassiterite fragments

(B) inner upper wall of crucible

Discussion:

The PhD survey project was undertaken to test and hopefully confirm the prehistoric use of cassiterite ore at sites in the Mid-Central Iberian region. References to the Mid-Central Iberian cassiterite ore areas are not well documented or extensively used in current historical or archaeological literature.

Modern geological literature (mostly of Spanish origin) documents that cassiterite ore locations are quite extensive, but the literature leaves out many of the older and smaller worked cassiterite ore sites.

The Mid Central Iberian tin research project consisted of field surface survey work and laboratory analysis of selected collected survey samples (mineral, metal and slag), plus the dating whenever possible of any surface collected pottery and flint.

The laboratory analyses helped confirm the identification of collected mineral, slag and metal samples.

One problem which became apparent during the early field survey work was the local population's general ignorance of former mine areas: local people generally had little or no knowledge of mine locations, ores mined and working dates of neighbouring mines.

The majority of the larger cassiterite and cassiterite/wolframite mines in Mid-Central Iberia ceased working quite recently, some as late as the mid 1980's. A majority of the older and smaller mines ceased production in the early to middle part of this century.

As time passes any undocumented or unlocated cassiterite ore sites will be even harder to identify as gradual erosion and vegetation covers the cassiterite ore areas.

It was soon apparent that the survey work would have to become long term if it was to be successful in gathering all available field site information.

It was important first and foremost to locate and document modern cassiterite ore sites, note the geography and geology of the site and determine whether the sites would support the criteria for being a possible, prehistoric archaeometallurgical site. The major criteria were considered to be a mineral site that would have been visible and accessible in antiquity and easily mined using prehistoric mining techniques for either vein or alluvial ore workings.

To date my survey work in Western Mid-Central Iberia suggests that at least one cassiterite site (El Cerro De San Cristobal, Visit No 4, Logrosan, pages 51-66) was providing local cassiterite ores for use in bronze production and for possible export during the Late Bronze Age. Any earlier dates than the Late Bronze Age for the mining and use the of cassiterite ores at El Cerro De San Cristobal will have to wait for future archaeological excavations.

At present El Cerro de San Cristobal has been dated solely by the surface collected pottery, the majority of which is Late Bronze Age. The collected stone hammers at the site can not at present be used for dating. It will be necessary to have excavated stratified stone hammers present with datable pottery or charcoal before the San Cristobal stone hammers and crushers can be used for definite dating purposes.

The archaeological artefacts, stone, flint, metal and pottery collected by Sos Baynat at El Cerro de San Cristobal point to an earlier archaeological date, possibly Copper Age (Appendix I & II).

The artefacts Sos Baynat collected from El Cerro De San Cristobal possibly indicate an earlier settlement site, which was not involved with cassiterite ore production. There does exist the possibility that gold or copper were being mined or worked at El Cerro de San Cristobal during the earlier Copper Age period.

To date no surface slag, furnace or crucible remains have been either found or identified at El Cerro de San Cristobal during the survey work.

The three metal artefacts from El Cerro de San Cristobal have analysed as Bronze (Sample No's 44, No 97 and No 98, page 135). To date no pure copper artefacts have been found.

The site at El Berrocal (Visit No 42, page 123-131), both settlement and mine areas have produced pottery, flint and stone tools. No furnace, crucible or slag remains have yet been found at El Berrocal. El Berrocal is at present dated by the pottery from the settlement site to the Copper Age.

In the Copper Age El Berrocal could have been a settlement site only without mining taking place though it is possible that the mining or working of gold or copper could have been undertaken during the Copper Age.

It is possible that the Cassiterite ore of the Berrocal site was mined during the Late Bronze age, but to date no archaeological evidence has been found definitely linking the El Berrocal site to cassiterite ore production.

The Berrocal site mine area needs more survey work and some small excavation trenches within the settlement area and in the mine area where the stone hammers were found in 1995.

The slags which were found at the sites of La Mina De Tres Arroyos (pages 69-70), The Mine at Santa Eulalia (pages 100-122), Minas Valdeflorez (pages 105-108), Area Catraia Do Buraco (pages 109-111) and Castelo Belmonte (page 115) were all identified as iron smelting or forging slags.

The lack of any quantity of early tin slag expect for the small amount of unique Roman tin slag from Torre Romana Centumcellas, at Belmonte, Portugal, (pages 92-94) still leaves unanswered the question of earlier cassiterite ore production and its use in the Mid-Central Iberian region.

The Roman tin slag from Torre Romana Centumcellas, Zezere valley, Portugal with the inclusion of the elements of Ti (titanium), Ta (tantalum) and Nb (niobium), (analysis results, (pages 144-145), clearly indicates that the slag was most probably derived from the use of local cassiterite ores of the Zezere Valley region.

The Zezere valley slag appeared in a small amount. Normally when slag is being produced during a metallurgical process such as smelting the amount generated exists in abundance. What is lacking at Torre Romana Centumcellas is even the evidence of crushed slag fragments, where the remaining tin prills in the slag would have been removed and further refined.

The direct addition method of cassiterite mineral to refined copper would have been the most straightforward way to produce bronze in pre history.

The general lack of tin slag during my field survey work must help to support the argument that only rich cassiterite mineral was being smelted into tin ingots or alternatively that sorted rich cassiterite mineral was being added directly to refined melted copper to make bronze. Neither of these two metallurgical process would produce large amounts of slag, if any at all.

The Vilaca crucible fragments are rim fragments only. It is probable that the bottom parts of the crucibles which would not have been as vitrified or slagged as the upper rim area, were broken up and destroyed during or after the bronze making process.

The microprobe analytical results from Raquel Vilaca's Portuguese crucible fragments (pages 155-160) seem to show that two of her excavated Late Bronze Age settlement sites in the Central Mid-Iberian region were using processed cassiterite ore directly for the bronze making process, after crushing the cassiterite and sorting out the accompanying gangue material. The finely crushed cassiterite with accompanying covering charcoal was then being added directly to pre melted refined copper in a crucible operation to produce bronze for casting purposes.

The appearance of metallic prills of pure copper and bronze prills containing varying percentages of tin within the inner wall of the crucible fragments along with what seems to be embedded crushed cassiterite ore fragments, seems to suggest that the crucibles were being used for bronze production. The embedded cassiterite fragments are only to be found in the inner top area wall of the rim crucible fragments, fragmented cassiterite ore was not found in the general ceramic matrix or outer walls of the crucible.

Gerry McDonnell of the University of Bradford has suggested in a useful discussion the possibility that the embedded cassiterite fragments might be the result of tin metal diffusion as vapour into the crucible walls and then re-oxidation of the tin metal to a tin oxide.

It should be noted however that the boiling point of tin is 2269.85 Celsius (^{0}C), and that tin would probably sublime at a temperature in the region of 1800 Celsius(^{0}C), this quite high temperature range would probably not have been achievable in a pre-historic crucible process.

The existence of what seem to be very fine slender prismatic shaped cassiterite crystals embedded in the inner crucible walls strongly suggests that the cassiterite present in the cracked, fragmented, vitrified crucible upper inner walls is still in its natural unaltered form. The microprobe analysis of these cassiterite crystals fragments seem to support this conclusion (Sample, Cast No 11, Areas 1-4, page 157).

Another question which arises is just how did the unaltered very finely crushed cassiterite ore fragments get into the crucible walls. One possible answer could be that the crucible walls in the area just below the crucible rim where the cassiterite would have been placed on top of the molten copper had been badly fractured or fragmented either during a preheating process of the crucible or during the preliminary process of presmelting of the refined copper, when the crushed cassiterite ore was put into the crucible the fragmented or fractured crucible walls allowed for the influx or embedding of some of the finer fragments of the cassiterite ore before it was able to smelt into tin metal and mix with the molten copper.

If the Vilaca crucibles were being used exclusively to smelt crushed cassiterite into tin, then the cooper prills (Sample, Cast No 11, Areas 5 & 6, page 157).and bronze prills (Sample, Ale No 37/B, Area 1, page 157) would not have been present in the crucible walls mixed with and below the finely embedded crushed cassiterite fragments

Simon Timberlake in a bowl furnace tin smelting experiment carried out at Flag Fen, May 1994 seems to have had also the embedding or entrapment of tin prills in furnace clinker and adhering clay furnace lining. In his report Simon Timberlake did not go into detail of the causes of the alteration of the bowl furnace lining and the subsequent entrapment of the tin within the furnace wall clinker and clay (Timberlake, 1994, pp 121-128).

It could be noted that it is possible the Vilaca crucibles had been re-used for cassiterite ore smelting after they had been used first in bronze production or in the reverse order, but this suggestion would be difficult to prove. It is doubtful that the crucibles used in prehistoric metallurgy work would have been used more than once due to the destructive nature of the crucible metallurgical process.

The almost complete lack of tin slag in the ancient world suggests that the addition of unsmelted cassiterite to refined copper seems to have been the primary process for using tin to make bronze. Evidence though from only two crucible fragments is not enough to make any final judgement on this important pre historical metallurgical question.

Further analytical work needs to be done on the Vilaca crucible fragments to try and answer all the questions that have arisen from the original analysis results; analysis of the ceramic matrix and further analysis in the areas around where the crushed cassiterite ore fragments are located.

It is necessary in the future to do a set of back scattered electron photographs through the crucible wall, from the inner wall to the outer wall. The back scatter photographs with accompanying analysis should clarify if cassiterite ore is present through out the fabric of the crucible or only on the upper inner surface area.

On the question of the movement of tin as a vapour (as suggested by Gerry McDonnell), some future laboratory experimentation needs to be done to see if the diffusion of the tin metal as vapour and the resulting re-oxidation of the metal to tin is possible within a ceramic crucible matrix.

The tin/copper mineral which I found at the Golpejas mine site and which I was able to smelt into a natural bronze in a small ceramic furnace adds another interesting aspect to the development of early bronze use by prehistoric man. I have been told that different variations of tin/copper minerals were present in quite a large area of Central Iberia, but that most of these minerals have by now been mined out (Felix Garcia Palomero and Ignacio Montero Ruiz, personal communications, 1993 & 1994). Being able to prove that prehistoric man used this complex mineral to produce a natural bronze remains almost impossible.

The area surrounding the smelted copper tin bronze prills in the Golpejas copper tin furnace smelted mineral samples (Sample No's 74 & 82, page 140) could do with further analysis to help give a more complete understanding of the smelting process that took place in the copper tin mineral.

I had hoped that the green bead fragment, very similar in visual colour and texture to the Golpejas copper tin mineral, which had been excavated from the El Torrejon megalithic monument near Golpejas would prove to be of the same elemental makeup as the Golpejas copper tin mineral. If the bead was of the same elemental make up this would indicate that prehistoric man during the Neolithic period was at least aware of the very beautiful green coloured copper tin mineral which was to be found in the Golpejas area and that he had used it to fabricate objects of adornment.

The non destructive surface analysis of four areas of the El Torrejon bead fragment (SEM/EDS) analysis (Sample No 108) revealed no copper or tin present. The elements present were Aluminium (Al) Phosphorus (P), Silica (Si) and small traces of Potassium (K), Calcium (C) and Iron (Fe).

I have included the bead fragment analysis here as the laboratory work was not finished early enough to include the analysis in my analytical chapter. All analyses are given in wt%.

Golpejas		%	%	%	%	%	%	%	%	%	%	%	%	%	%	%	%
Megalitico de		Sn	Cu	Ag	Ti	Ta	Al	Si	Fe	Mn	S	K	P	Ca	Cl	Hg	TOTAL
"El Torrejon"																	
Visit 16																	
Fragment	Area 1				0.10		11.6	14.01	2.95		0.13	2.57	6.13	0.34			37.82
de Colgante	Area 2						7.66	0.26	1.16			0.21	13.8	0.10	0.14		23.41
4.12.95	Area 3				0.10		5.20	1.37	1.39		0.13	0.35	8.08			0.52	17.23
	Area 4						2.40	6.14	3.78		0.10	0.72	0.26	0.29	0.22		14.08

Table No 18 SEM/EDS analysis, Fragment de Colgante, Sample No 108

Some of the most promising geological tin mining sites which I surveyed yielded little or no ancient remains. These sites certainly deserve more survey work.

The following survey report chart (pages 164-165) gives a quick synopsis of the sites which I visited and the results and whether or not more time should be spent surveying some of the sites.

SITE	LITERARY REFERENCE	SURVEY RESULTS	COMMENT
Visit 1			
Mina Mari Juli Y Mina Virgin De Garcia	none	no evidence	**further work not recommended**
Visit 2			
Groupe Mina San Nicholas	none	no evidence	further survey required
Visit 3			
Mina La Parrilla	none	no evidence	further survey required
Visit 4			
El Cerro De San Cristobal-Logrosan	yes	Late Bronze Age	further survey and excavation work required
Visit 5			
La Mina Plasenzuela	yes	no evidence	further survey required
Visit 6			
La Mina de Tres Arroyos	yes	Roman	further survey required
Visit 7			
Mina de Santa Maria	none	no evidence	**further work not recommended**
Visit 8			
Mina Calero	none	no evidence	**further work not recommended**
Visit 9			
Mina La Lapa	none	no evidence	**further work not recommended**
Visit 10			
La Mina Desquites	none	evidence found no date	further survey required
Visit 11			
Mina Comohal Hubaraba Y Horrera	none	evidence found no date	further survey required
Visit 12			
La Mina Carrascal	none	no evidence	**further work not recommended**
Visit 13			
La Mina Montehermoso	none	no evidence	**further work not recommended**
Visit 14			
Mina Telba	yes	evidence found no date	further survey required
Visit 15			
Mina Claudina	none	no evidence	**further work not recommended**
Visit 16			
Mina Golpejas	yes	no evidence	further survey required
Visit 17			
Mina Barruecopardo	none	no evidence	**further work not recommended**
Visit 18			
Site of Les Merchanas	yes	no evidence	further survey required
Visit 19			
Mine at Barquilla	none	no evidence	**further work not recommended**
Visit 20			
Torre Romana Centumcellas	yes	Roman	further survey required
Visit 21			
Mountain of Fraguas	none	no evidence	**further work not recommended**

Table No 19 Survey Sites - Evaluation List

SITE	LITERARY REFERENCE	SURVEY RESULTS	COMMENT
Visit 22			
The Lajes Mine	none	no evidence	**further work not recommended**
Visit 23			
The Gois Mine	none	no evidence	further survey required
Visit 24			
The Panasqueira Mine	none	no evidence	**further work not required**
Visit 25			
The Santa Eulalia Mine	none	evidence found	**further work not required**
Visit 26			
Ruinas Romanas de Sao Cucufate	none	evidence found	copper area
		no date	
Visit 27			
The Ajustrel Mine	yes	evidence found	copper area
		no date	
Visit 28			
Casas de Don Antonio	none	no evidence	further survey required
Visit 29			
Minas Valdeflorez	yes	evidence found	further survey required
		no date	
Visit 30			
Area Catraia Do Buraco - Belmonte	none	evidence found	further survey required
		no date	
Visit 31			
Search for Santa Maria mine	none	no evidence	further survey required
Visit 32			
Bronze Age Site Hill 649 & Hill 648	none	no evidence	**further work not recommended**
above Quinta da Chandeirinha			
Visit 33			
Ribera Bendada - south of village Olas	none	no evidence	**further work not recommended**
Visit 34			
Area South East of Belmonte - Gare	none	no evidence	**further work not recommended**
Visit 35			
Ponte do Belmonte	yes	no evidence	further survey required
Visit 36			
Castelo Belmonte	none	evidence found	further survey required
		no date	
Visit 37			
Povado de Misericordia - Serpa	none	evidence found	copper area
		no date	
Visit 38			
Mina de San Expedito	none	evidence found	further survey required
		no date	
Visit 39			
Mina El Terruco	none	no evidence	further survey required
Visit 40			
Mina La Periza - Montanchez	none	no evidence	further survey required
Visit 41			
Minas Doninos de Ledesma	none	no evidence	further survey required
Visit 42			
Mina de Berrocal - Merdia	none	**Copper Age**	further survey and excavation work required

Table No 19 Survey Sites - Evaluation List

Further Work:

Of the mines listed above for further survey work the following observations apply:

Mina San Nicholas, Visit 2, cassiterite mineralization probably visible on surface during antiquity.

El Cerro de San Cristobal, Logrosan, Visit 4, further planning of site and worked out cassiterite veins and small excavation trenches.

La Mina de Tres Minas, Visit 6, remains of Roman furnaces, possibly tin smelting (?) possible Roman pottery found, (Fe) iron slag found.

L Mina Desquites, Visit 10, easily accessible cassiterite in antiquity due to the kaolinasation of part of the granite mineral bearing deposit.

Mina Telba, Visit 14, easily accessible cassiterite surface ore in some small areas of mine, possible pre-historic pottery and stone tool fragment found.

Mina Golpejas, Visit 16, location of copper/tin ore and mention in some literature sources as possible Late Bronze age mine site.

Torre Romana Centumcellas, Visit 20, only site where tin slag was found.

The Santa Eulalia Mine, Visit 25, easily accessible cassiterite ore, possible Roman pottery found.

Minas Valdeflorez, Visit 29, cassiterite ore veins visible on surface in antiquity and mention in literature source as a site of Roman tin workings.

Area Catraia do Buraco, Visit 30, iron slag found with possible Late Bronze Age pottery and stone tools.

Mina de San Expedito, Visit 38, cassiterite ore visible surface in antiquity and easily accessible, possible pre-historic pottery found.

Mina El Terruco, Visit 39, actual mine still needs to be located.

Mina La Periza-Montanchez, Visit 40, further preliminary survey work of this area still remains to be done.

Minas Doninos de Ledesma, Visit 41, further preliminary survey work needs to be done in this area.

Mina El Berrocal, Visit 42, prehistoric pottery and stone hammers and grinders found at this site, cassiterite ore visible and easily accessible in antiquity, more survey of mine area needed and small excavation trenches for stratigraphy needed.

Conclusions:

The survey research project has generated many new questions for a more advanced level of investigation. Exactly how many tin mining areas had been worked in the Mid Central Iberian region? Will it be possible in the future to get some idea of the amount of tin that has been mined during antiquity within the Mid-Central Iberian geological region? Will there be proof forthcoming that tin was not only mined and used locally, but also traded out to the Western Atlantic coastlands and to the Southern Atlantic and Mediterranean coasts?

Small archaeological excavation trenches will be needed at El Cerro de San Cristobal and El Berrocal to provide precise archaeological stratigraphy of the mining and settlement areas and to definitely confirm the use of local cassiterite ores in the Late Bronze Age and possible earlier use.

The mined out cassiterite veins at El Cerro De San Cristobal which are quite extensive will need to be properly surveyed and planned to get some sort of understanding as to the actual amount of cassiterite ore mined in the prehistoric period.

This survey has not proved conclusive evidence of Late Bronze age tin mining and smelting, but has offered some very strong preliminary proof that local cassiterite ore was being mined and smelted at this date. I remain convinced that with further field survey work more evidence of other prehistoric cassiterite mineral sites in the Mid-Central Iberian region will emerge.

Also including selected references not cited in the text.

Ager, D.V., 1980
The Geology of Europe
London: England: McGraw Hill

Agricola, G.
De Re Metallica
(translated from the first edition of 1556
by Hoover H.C.;Hoover L.H., 1950)
New York: Dover Publications.

Alarcao, J.de., 1988
Roman Portugal, Volume I & Volume II Fascicule 1-3
Warminster: England: Aris & Phillips.

Arribas, A., 1963
The Iberians
London: England: Thames and Hudson.

Atienza, F.B., 1988
Nuevo Diccionario Politecnico De Las Lenguas Espanol E Ingles
Nuevo Diccopnario Politecnico De Las Lenguas Ingles E Espanol
Volumen Primero - Volumen Segundo
Madrid: Spain: DIAZ DE SANTOS, S.A.

Atkinson, R.L., 1985
Tin and Tin Mining Shire Album 139
Aylesbury: England: Shire Publications.

Aubet, M.E., 1993
The Phoenicians and the West
Cambridge: Cambridge University Press.

Bachmann, H.G., 1982
The Identification of Slags From Archaeological Sites
Institute of Archaeology - Occasional Publication no 6
London: University College London.

Barrena, M.A. & Isidro, P.D., 1987
La Minera En Extremadura
Merida: Spain: Junta De Extremadura.

Bass, G.F., 1987
Oldest Known Shipwreck
National Geographic Vol 172 No 6
Washington D. C., National Geographic Society.

Bates, R.L. and Jackson, J.A., 1990
Glossary Of Geology Third Edition
Alexander: Virginia: American Geological Institute.

Beagrie, N., 1985
Some early tin ingots, ores and slags from Western Europe
The Journal of the Historical Metallurgy Society,
Volume19 no 2 ; London: Institute of Metals.

Ben-Tor, A., 1992
The Archaeology of Ancient Israel
London: Yale University Press.

Berrocal-Rangel, L.; Enriquez Navascues, J.J.; Celestino Perez, S.; Valdes Fernandez, F., (Coordinacion), 1995
Extremadura Arqueologica IV
Arqueologia En Extremadura:10 Annos De Descubrimientos
Merdia: Spain, Junta De Extremadura.

Bousfield, B., 1992
Surface Preparation and Microscopy of Materials
Chichester: John Wiley & Sons.

Bowman, S., (ed) 1991
Science and The Past
London: British Museum Press.

Brant, D., 1985
Metallurgy fundamentals
South Holland, Illinois: Goodheart-Willcox.

Britton, D., 1963
Traditions of Early Metal-Working in the later Neolithic and Early Bronze Age of Britian Part 1
The Prehistoric Society No 10
London: The Prehistoric Society.

Brooke, J., 1994
The Tin Streams of Wendon
Turo: Cornwall: Twelveheads Press.

Brown, D V., 1981
Basic Metallurgy
Albany: New York: Delmar Publishers.

Bygott, J., 1938
An Introduction To Mapwork and Practical Geography
London: University Tutorial Press ltd.

Chapman, R., 1990
Emerging complexity - The later prehistory of south-east Spain, Iberia and the west Mediterranean
Cambridge: Cambridge University Press.

Clark, J.G.D., 1952
Prehistoric Europe
London: Methuen & Co Ltd.

Clayton, B., 1974
Metal Analyyses-their limitations and applications to the Early Bronze Age in Ireland - Bulletin No 11
London: Institute of Archaeology.

Cleuziou, S., and Berthoudt, T., 1982
Early Tin in The Near East; Expedition, 25: 91 pp 14-25.

Coles, J.M. and Harding, A.F., 1979
The Bronze Age in Europe; London: Methuen and Co.

Consejeria de Economia y Hacienda/ Castilla Y Leon 1988
Los Recursos Minerales de Castilla y Leon NO 2 Estano
Valladolid: Spain: Junta de Castilla y Leon.

Also including selected references not cited in the text.

Cook, D. and Kirk, W., 1991
Field Guide to Rocks and Minerals of the World
London; Kingfisher Books, Grisewood and Dempsey.

Costa, E H.M., Dos Santos, J.R., 1965-1966
Duas Campanhas de Excavacoes No Castro de Carvalhelhos
Trabalhos de Anthropologie e Etnologia
Volume XX, Faces 1-2 pp 181-190: Porto: Portugal.

Craddock, P.T., (ed)., 1980
Scientific Studies in Early Mining and Extractive Metallurgy
British Museum Occasional Paper No 20
London: British Museum.

Craddock, P.T. and Hughes, M.J.,(ed), 1985
Furnaces and Smelting Technology in Antiquity
British Museum Occasional Paper No 48
London: British Museum.

Craddock, P.T., (ed), 1990
2000 Years of Zinc and Brass
British Museum Occasional Paper No 50
London, England; British Museum Publications.

Craddock, P.T., 1995
Early Metal and Mining Production
Edinburgh, Sctoland; Edinburgh University Press.

Cronyn, J. M., 1990
The Elements Of Archaeological Conservation
London: Routledge.

Cuncliffe, B. and Keay, S.,(ed), 1995
Social Complexity and the Development of Towns in Iberia
From The Copper Age To The Sevcond Century A.D.
Oxford: Oxford University Press.

Dallmeyer, R.D.and Garcia, M.E., (ed), 1990
Pre-Mesozoic Geology of Iberia
Berlin: Springer-Verlag.

Davies, G.M., 1919
Tin Ores; London: John Murry-Albemarle Street, West.

Davies, O., 1935
Roman Mines in Europe; Oxford: Claredon Press.

Day, J. and Tylecote, R.F., (ed), 1991
The Industrial Revolution In Metals
London: The Institute of Metals.

Delibes, G and Santonja, M., 1986
El Fenomeno Megalitico En La Provincia De Salamanca
Salamanca, Spain: Ediciones de la Diputacion de Salamanca.

Dines, H.G., 1988
The metalliferous mining region of south-west England
Vol I & II: London: British Geological Survey.

Diodorus, Siculus
Diodorus Siculus III Books IV, 59- VIII
The Leob Classical Library-translated by C.H. Oldfather, 1939; London: William Heinemann ltd.

Domerque,.C., 1987
Catalogue Des Mines Et Des Fonderies Antiques
De La Peninsule Iberique-Tome I y Tome II
Madrid: Spain: Publications de la Casa de Velazquez
Serie Archeologie VIII-Diffusion de Boccard.

Domerque, C., 1990
Les Mines De La Peninsule Iberique Dans L'Antique Romaine Collection De L'Ecole francaise de Romaine-127
Roma, Italie: Scuola Tipografica S. Pio X.

Dos Santos, J.R., 1965-1966
Trabalhos de Antropopogia e Etnolgia 1965-1966
Tin Containing Slags at Carvalhelhos Portugal
Porto: Portugal, Sociedade de Portugese
Antropologia e Etnologia.

Dunning, F.W.; Garrard, P. ; Halsam, H.W.; Ixer, R.A., 1989
Mineral Deposits of Europe
Volume 4/5; Southwest and Eastern Europe with Iceland
London: the Institution of Mining and Metallurgy.

Duque, A.M. and Martinez, J.M.B., 1982
Espana Romana (218 a de J.C. - 414 de J.C.) Volumen 1
La Connquista y La Explotacion Economica
Madrid: Espana, Espasa-Calpe S.A.

Edmondson, J.C., 1987
Two Industries in Roman Lusitania/Mining and Garum Production; Oxford: BAR International Series 362.

Elphick, G., 1988
The Craft Of The Bell Founder
Chischester: Phillimore & Co.

Embleton, C., (ed), 1984
Geomorphology of Europe
London: Macmillan Publishers.

Embry, P.G. and Symes, R.F., 1987
Minerals of Cornwall and Devon
London: British Museum.

Emsley, J., 1989
The Elements; Oxford: Clarendon Press.

Forbes, R.J., 1972
Studies in Ancient Technology-Volume IX
Leiden: Netherlands, E.J. Brill.

Franklin, A.D.; Olion, J.S.; Wertime, T.A., 1978
The Search For Ancient Tin
Washington D.C.: Government Printing Office.

Also including selected references not cited in the text.

Galan, M.B., (Coodinador), 1991
Historia General De Espana Y America
De La Protohistoria A La Conquista Romana
Tomo I-2; Segunda Edicion
Madrid: Espana, Ediciones Rialp S.A.

Gallo, N.G., (Coordinador), 1985
Historia General De Espana Y America
Los Origenes De Espana-Tomo I-1
Madrid: Espana, Ediciones Rialp S.A.

Goodhew, P.J., 1975
Electron Microscopy and Analysis
London: Wykeham Publications.

Graham-Campbell, J., (ed), 1988
World Archaeology Volume 20; No.1; June 1988
London: Routledge.

Grant, M., 1980
The Etruscans; London: Weidenfield and Nicolson.

Greeves, T., 1986
Tin Mines and Miners of Dartmoor - A Photographic Record
Devon: Devon Books.

Guido, M., 1963
Sardina; London: Thames and Hudson.

Gumel, P., Novembre-Decembre 1981
Chronique De La Recherche Miniere
Essai sur la Classification Typologique des Principaux Gisements de Sn-W Estremadure (Espange)
Paris, France, Centre D' Etudes Geologiques.

Guzman, F.V. 1983
Depositos Minerales de Espana
Madrid: Espana: Instituto Geologico y Minero de Espana

Harden, D., 1962
The Phoenicians; London: Thames and Hudson.

Harris, D., (ed), 1989
Cornish Archaeology No.28
Cornwall: Cornwall Archaeological Society.

Harrison, R.J. 1977
The Bell Beaker Cultures Of Spain And Portugal
Cambridge: Massachusetts: Peabody Museum.

Harrison, R.J., 1980
The Beaker Folk-Copper Age Archaeology in Western Europe; London: Thames and Hudson.

Harrison, R.J., 1988
Spain at the Dawn of History-Iberians, Phoenicians and Greeks; London: Thames and Hudson.

Hodges, H., 1964
Artefacts; London: John Baker Publishers.

Instituto Geologico y Minero de Espana, 1988
Monografias De Sustancias Minerales-Estano
Madrid: Servicio De Publicaciones Ministerio De Industria.

Jimenez, A.V. and de la Barrera Anton, J.L., (Coordinacion), 1993
El Proceso Historico De La Lusitania Oriental
En Epoca Preromana Y Romana Cuadernos Emeritenses-7
Merdia: Espana, Museo Nacional De Arte Romano.

Jorda Cerda, F.; Pellicer Catalan, M.; Acosta Martinez, P.; Almagro-Gorbea, M., 1989
Historia De Espana Prehistoria 1
Madrid: Espana, Editorial Gredos S.A.

Junta De Extremadura, 1993
Minerales Gemologicos De Extremadura
Merida: Espana, Consejeria de Industria y Turismo.

Junta de Castilla Y Leon, 1988
Los Recursos Minerales de Castilla y Leon No 2 Estano
Valladolid: Espana, Consejeria de Economia y Hacienda.

Keay, S.J., 1988
Roman Spain
London: British Museum Publications Limited.

LaLaguna, J., 1990
A Traveller's History of Spain
Moreton-in-Marsh: U.K., The Windrushpress.

Lehmann, B., 1990
Metallogeny of Tin/Lecture Notes in Earth Sciences 32
Berlin: Springer-Verlag.

Lourenco, C.;Goinhas, J.;Martins, L.;Viegas, L.;Gonzalez, V., 1992
Portugal-Exploration And Mining: An Interesting Challenge
Lisbon: Portugal, Servico De Fomento Minero.

Maddin, R., (ed), 1988
The Beginning of the Use of Metals and Alloys
London: The Massachusetts Institute of Technology.

Madronero De La Cal, A., 1994
Una Posibilidad De Rastreo De Los Origenes De La Metalurgia Del Estano En Espana
Coruna, Espana: Edicios Do Castro.

Ministry of Defence, 1988
Manual Of Map Reading And Land Navigation
London: HMSO.

Molero, C.A., 1993
La Minera En Extremadura
Merida: Spain, Junta De Extremadura.

Also including selected references not cited in the text.

Montenegro, A., Blazquez, J.M., Mata, D.R. , Garcia, C. J.M., Iniesta, A., Fatas, G., Salinas, M., Pastor, Y.M., 1989
Historia De Espana Colonizaciones Y Formation De Los Pueblos Preromanos (1200-218 a.C.) 2
Madrid, Espana: Editorial Gredos S.A.

Montenegro Duque, A.; Blazquez Martinez, J. M.; Solana Sainz Jose, Ma., 1986
Historia De Espana Espana Romana 3
Madrid: Espana, Editorial Gredos S.A.

Muhly, J.D. 1973
Copper and Tin; Hamden: Connecticut, Archon Books.

Muhly, J.D. 1976
Supplement To Copper and Tin
Hamden, Connecticut: Archon Books.

Navascues, J.J.E. and Diaz, A R., (Edicion), 1991
Extremadura Arqueologica II 1991
I Jornadas De Prehistoria Y Arqueologia Ee Extremadura (1986-1990)
Merdia: Espana, Consejeria de Educacion Y Cultura De La Junta De Extremadura, Universidad De Extremadura.

Nichel, E.H. and Nichols, M.C., 1991
Mineral Reference Manual
New York: Van Nostrand Reinhold.

Nicolini, G., 1973
The Ancient Spainards
Farnborough: England, Saxon House.

Ochoa, C.Q., 1987
Mapa Geologico De Extremadura
Merida: Spain, Junta De Extremadura.

Oddy, W.A., (ed), 1991
Aspects of Early Metallurgy
British Museum Occasional Paper 17
London: British Museum Occasional Papers.

Ortiz, A.D., (dirigida), 1990
Historia De Espana
desde la prehistoria hasta la conquista romana (siglo III a.C.)
Barcelonia: Espana, Editorial Planeta S.A.

Pardo, A., 1976
The World Of Ancient Spain
Geneve: Switzerland, Editions Minerva.

Parkes, P.A., 1986
Current Scientific Techniques in Archaeology
London: Croom Helm.

Pellant, C., 1992
Rocks And Minerals; London: Dorling Kindersley.

Penhallurick, R.D., 1986
Tin in Antiquity; London: Institute of Metals.

Piggott, S., 1983
The Earliest Wheeled Transport
From the Atlantic Coast to the Caspian Sea
London: Thames and Hudson.

Pitcher, W.S.,1993
The Nature and Origin of Granite
London: Blackie Academic and Professional.

Pliny,
Pliny Natural History II Books III - VII
Leob Classical Library-translated by H. Rackham, 1942
London: Harvard University Press.

Polunin, O. and Walters, M., 1985
A guide to the vegetation of Britian and Europe
Oxford: Oxford University Press.

Robertson, W., 1982
Tin: Its Production and Marketing; London: Croom Helm.

Roso De Luna, M., 1898, (1982 reprint)
Logrosan (Legajo Historico)
Caceres: Spain: Institucion Cultural "EL BROCENSE".

Rothenburg, B. and Blanco-Freijeiro, A., 1981
Ancient Mining and Metallurgy in South-West Spain
London: Institute for Archaeo-Metallurgical Studies.

Rothenberg, B., (ed), 1990
The Ancient Metallurgy of Copper
London: Institute for Archaeo-Metallurgical Studies.

Ruiz, I M., 1994
El Origen De La Metalurgia En El Sureste Peninsular
Almeria: Espana, Instituto de Estudios Almerienses.

Savory, H.N., 1968
Spain and Portugal-The prehistory Of The Iberian Peninsula
New York: Frederick Praeger.

Schuman, W., 1992
Minerals of the World; New York: Sterling Publishing.

Schumann, W., 1992
Rocks Minerals and Gemstones; London: Harper Collins.

Scott, D.A., 1991
Metallography and Microstructure of Ancient and Historic Metals
California: The Getty Conservation Institute.

Slater, D., 1974
Tin-Mineral dossier No 9
London: Her Majesty's Stationery Office.

Sos Baynat, V. 1965
Geologia de las inmediaciones de Merida (BADAJOZ)
Madrid: Extracto del "Boletin del Instituto Geologico y Minero de Espana" Tomo LXXV (1964).

Also including selected references not cited in the text.

Sos Baynat, V., 1967
Geologia, Mineralogia y Mineralogenia de la Sierra de San Cristobal, Logrosan (Caceres)
Madrid: Real Academia De Ciencias-Exactas, Fisicas y Naturales.

Sos Baynat, V., 1977
Los hallazgos prehistoricios de Logrosan (Caceres)
Badajoz, Spain: Diputacion Provincial De Badajoz.

Sos Bayant, V., 1977
Geologia De Los Alrededores De Casar De Caceres (Cacere.S)
Salamanca: Separata de Stvdia Geologica XII, pp 169-193.

Sos Bayant, V., 1980-1981
Sobre La Edad Geologica De Las Casiteritas De Extremadura, Espana
Lisboa: Sociedade Geologica De Portugal Boletim Volume XXII.

Sos Bayant, V, 1981
Los Yacimientos De Casiteritas Y Wolframitas De La Sierra De Montanchez, Caceres
Madrid: Bol. Soc. Esp., Mineralogia 3 (1981), p 5-14.

Sos Baynat, V., 1981
Museo De Geologia De Extremadura
Merida: Excmo Ayuntamiento de Merida.

Stambolov, T., 1985
The Corrosion And Conservation Of Metallic Antiques And Works Of Art
Amsterdam: Central Research Laboratoty for Objects of Art and Science

Stanier, R.P., 1988
Cornwalls Mining Heritage
Cornwall: Twelvehead Press.

Strabo,
The Geography Of Strabo II
Leob Classical Library-Translated by H.L.Jones, 1923
London: William Heinemann.

Tykot, R. H. and Andrews, T. K.,(ed), 1992
Sardina In The Mediterranean: A Footprint In The Sea
Sheffield: Scheffield Academic Press.

Timberlake, S., 1994. An Experimental Tin Smelt at Flag Fen.
The Journal of the Historical Metallurgy Society, 28 No 2: 122-128

Tylecote, R.F., 1962
Metallurgy in Archaeology
London: England, Edward Arnold (Publishers) ltd.

Tylecote, R.F., 1987
The early history of metallurgy in Europe
London: Longman Archaeology Series.

Tylecote, R.F., Photos, E., Earl, B., 1988
The composition of tin slags from the south-west of England
World Archaeology Volume 20 no 1 June 1988
London: Routledge.

Tylecote, R.F., 1992
A History of Metallurgy
London: Institute of Materials.

Velazquez, A., De La Barrera, J.L., Enriquez, J.J., (edicion), 1990
La Cultura Tartesica Y Extremadura
Merdia: Museo Nacional De Arte Romano.

Watt, A., 1982
Longman Illustrated Dictionary of Geology
Harlow: England, Longman Group.

Whitehouse, R., (ed), 1983
Macmillan Dictionary of Archaeology
London: The Macmillan Press.

Whitten, D.G.A. and Brooks, J.R.V., 1972
The Penguin Dictionary of Geology
London: Penguin Books.

APPENDIX I

DRAWINGS OF SURFACE ARCHAEOLOGICAL FINDS

from LOGROSAN
El Cerro De San Cristobal

collected by VICENTE SOS BAYNAT
during the 1950's and 1960's

drawn by JOSEFINA MOLINA GARCIA
Museo Nacional de Arte Romano, Merdia

The following twenty three pages of drawings show the artefacts from Logrosan, which were all collected by Vicente Sos Baynat during the 1950's and 1960's.

These artefacts are now stored in the Museo Nacional De Arte Romano, Merdia, Extremadura, Spain. None of the archaeological artefacts from Logrosan are on public display. I was able to view and photograph the Logrosan artefacts and have them drawn in 1994.

The artefacts from Logrosan seem to indicate a long period of prehistory at El Cerro de San Cristobal. The finds are both surface and excavated. The miners working at Logrosan in the 1950's and 1960's found quite a large percentage of the finds during their elluvial mine working on both the summit and slopes of El Cerro de San Cristobal (Sos Baynat, 1977, 261-62).

No stratified record exists of the mine working excavated artefacts other than a general description of the find area and general depth and in most case not even that. Quite a few of the artefacts showed up during the washing out of the cassiterite ore during the benefaction shaking table operation (Sos Baynat, 1977, 264-72).

The very rich variety of archaeological finds flint, stone, pottery and metal point to a long term prehistoric presence in the Logrosan area. Included in the finds are early stone tools and two definite Beaker artefacts a stone wrist guard and a metal Palmela point (Savory, 1968, 183). There are also present one or two artefacts which could represent the Roman period (Alonso Rodrigvez Diaz, personal communication, 1994).

The Sos Baynat archaeological collection from Logrosan seems to differs somewhat in dates from the bulk of the surface pottery from the survey work done on El Cerro de San Cristobal, which represent the Late Bronze Age (Alonso Rodrigvez Diaz, personal communication).

The Sos Baynat article on the archaeological finds of Logrosan also mentions necklaces and some small gold artefacts and fragments being present. The total weight of the gold being 20 grams (Sos Baynat, 1977, 274). Both the necklaces and gold seem to have been misplaced, as they are not included with the Logrosan collection at the Museo Nacional Arte Romano, Merdia; perhaps they might reappear again sometime in the future.

Sos Baynat also states in his article that the majority of the archaeological artefacts came from two locations one on the western summit area of El Cerro de San Cristobal and a second on the lower Northern slope called Huertos de San Martin. I have marked both these areas in blue on a 1898 light sketch of the Logrosan area from the publication by Roso De Luna (Roso de Luna, 1898, 40-1).

I was allowed by the National Roman Museum, Merdia to cut and drill small metallurgical samples from the ten metal objects from the Logrosan collection in the summer of 1995. I hope to analysis these metallurgical samples in the winter of 1995-1996. It will be interesting to see the matallurgical make up of the ten metal artefacts, whether they are copper, arsenical copper or bronze and if lead is present in any of the samples.

The results of the Logrosan metallurgical analysis are presented in Appendix II pages 194-203.

The drawing of the bronze chisel/drill (Sample No 97) found by Anthony Bridgeman in the settlement area of El Cerro de San Cristobal during the 1994 planning season has been placed beside the drawing of the casting mould from the Sos Baynat collection (page 194). The chisel fits precisely into the casting mould.

The scale of all the drawings by Josefina Molina Garcia are either one to one or one half of the actual artefact size.

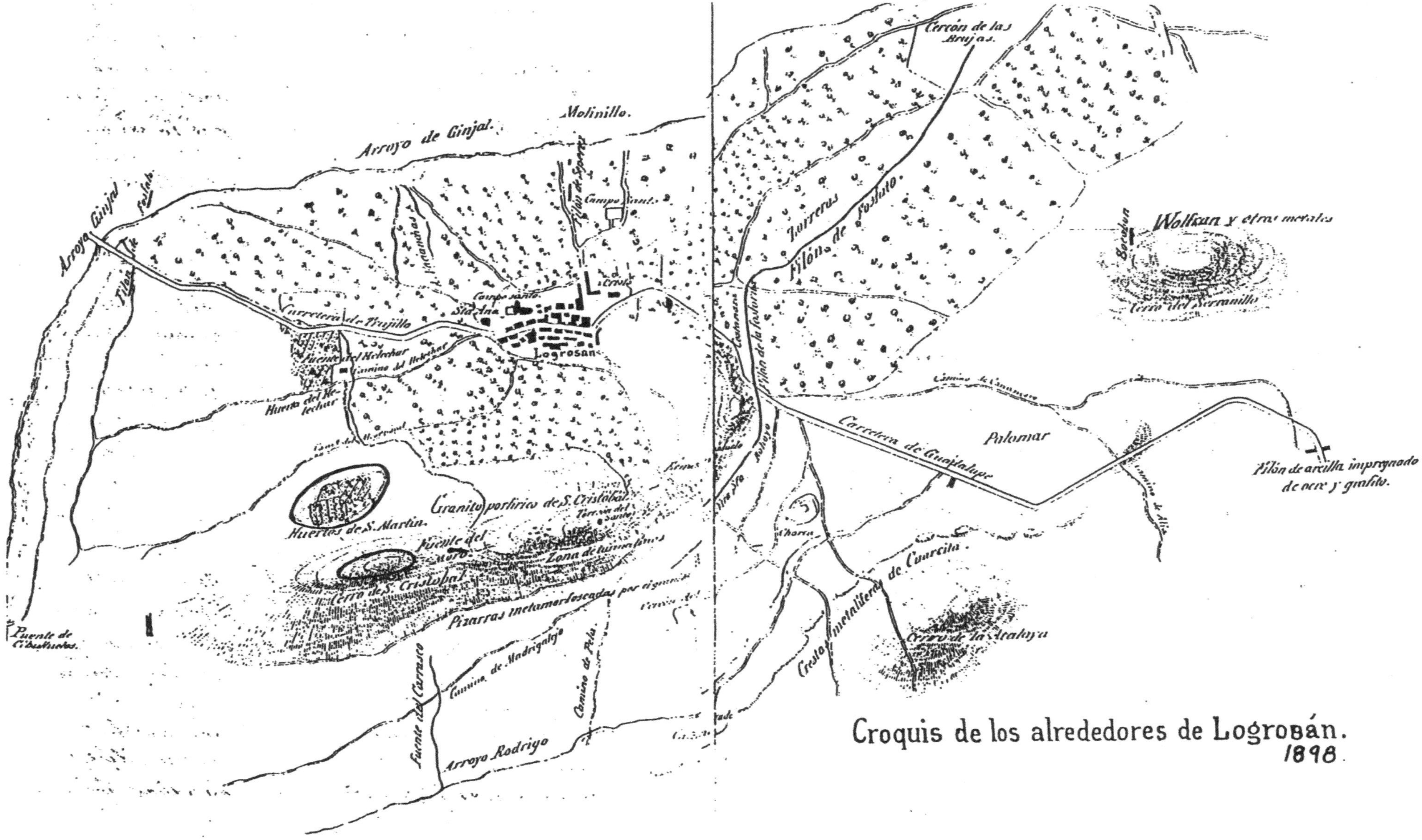

Croquis de los alrededores de Logrosán.
1898.

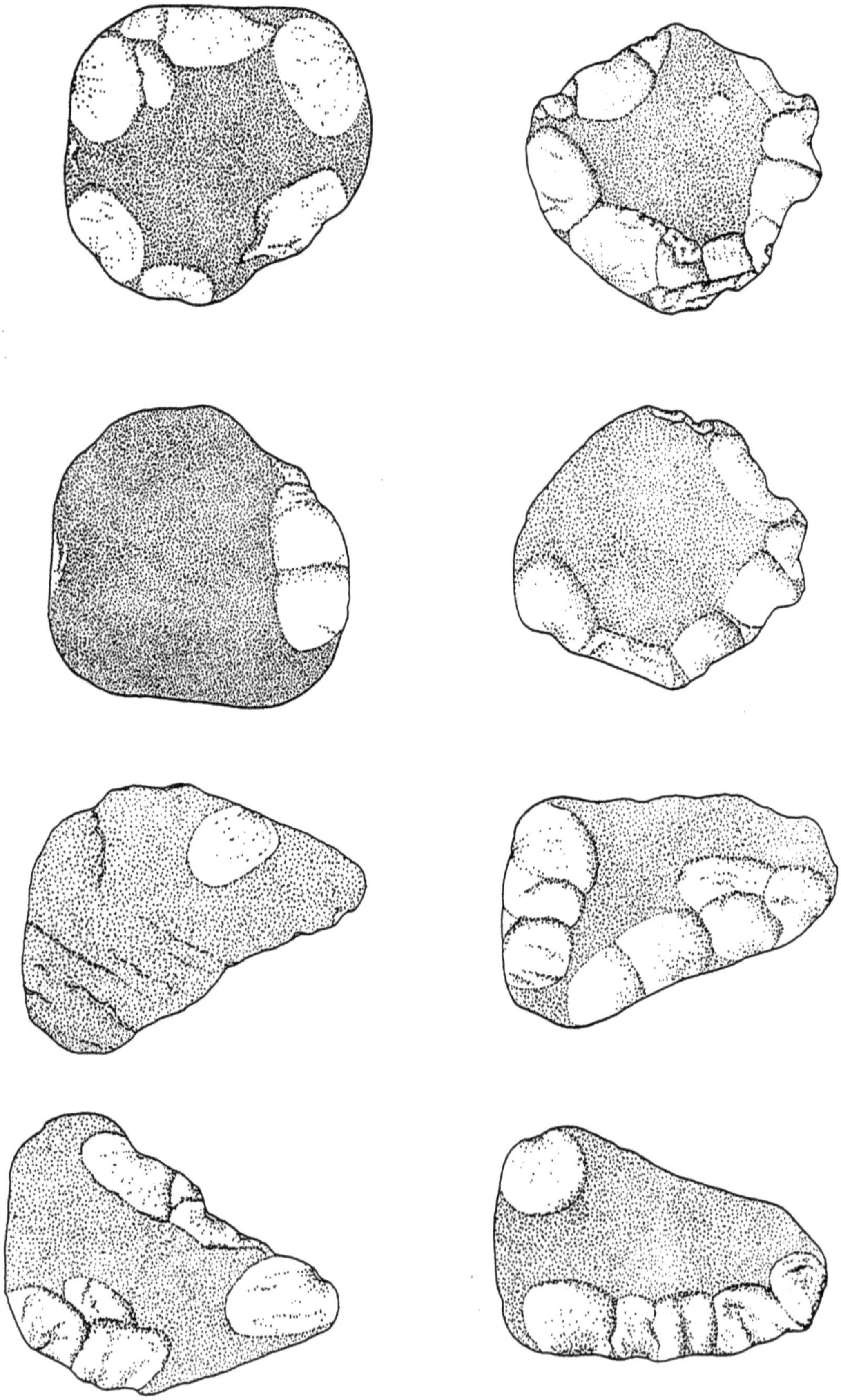

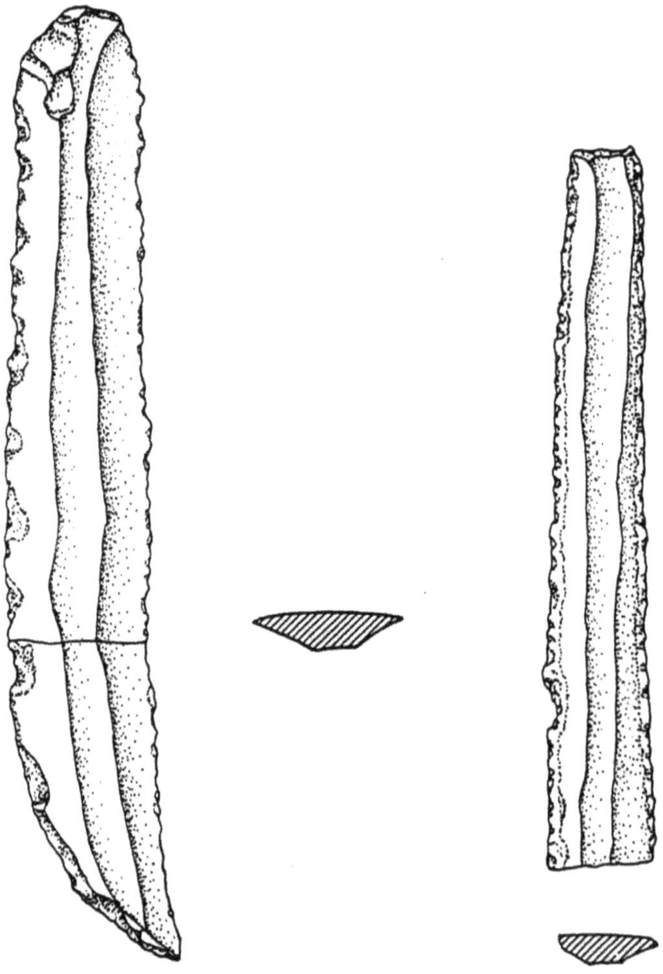

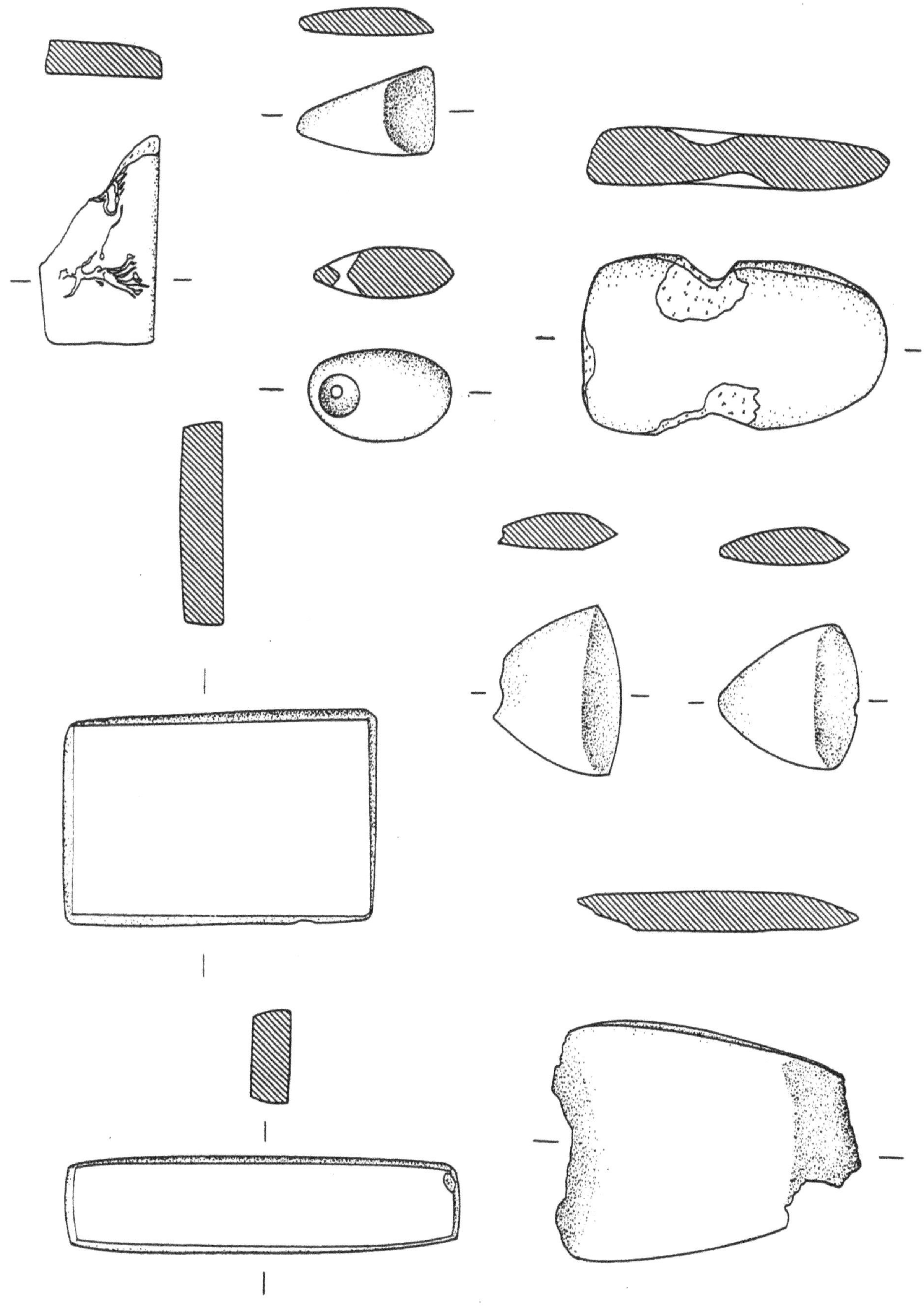

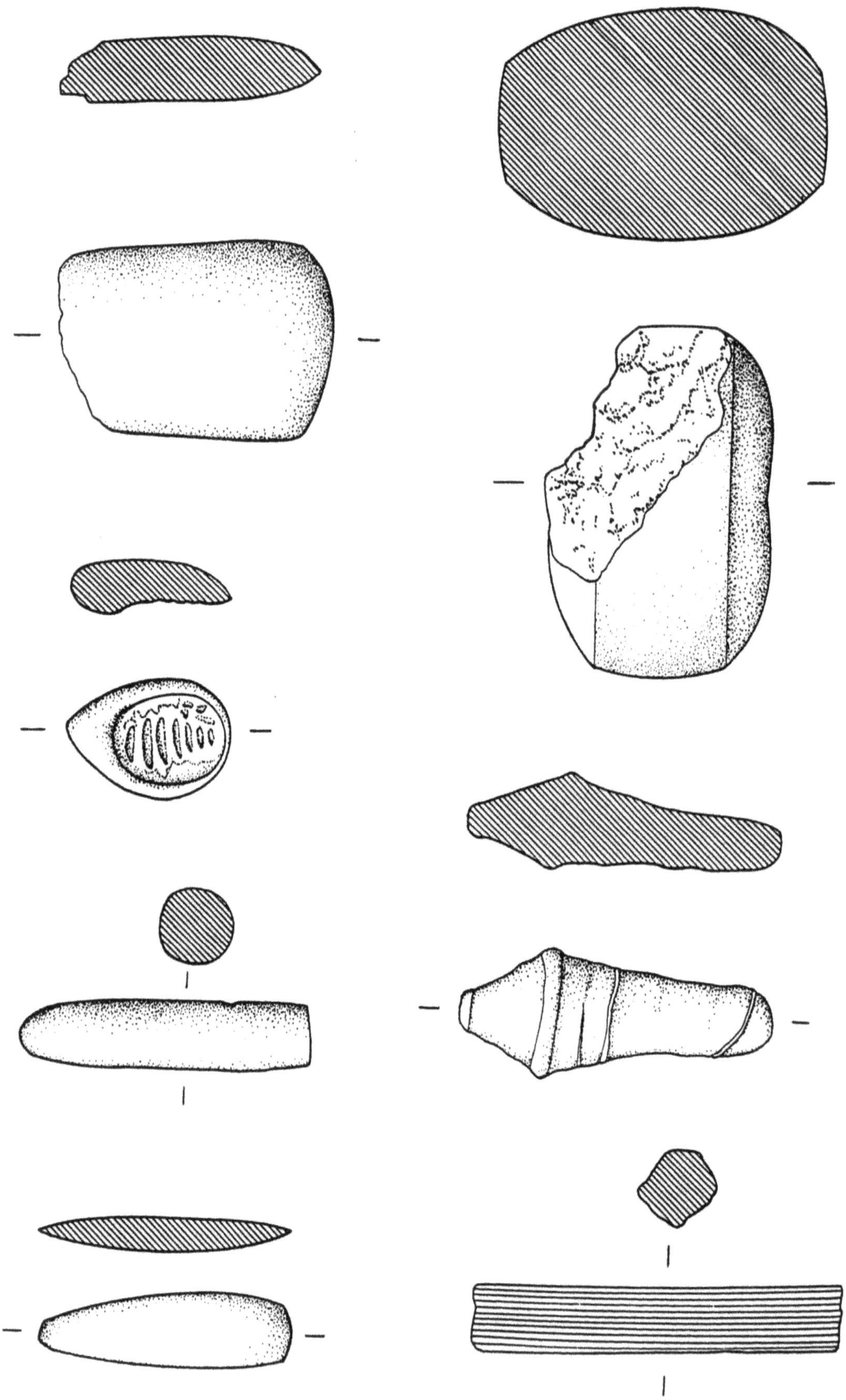

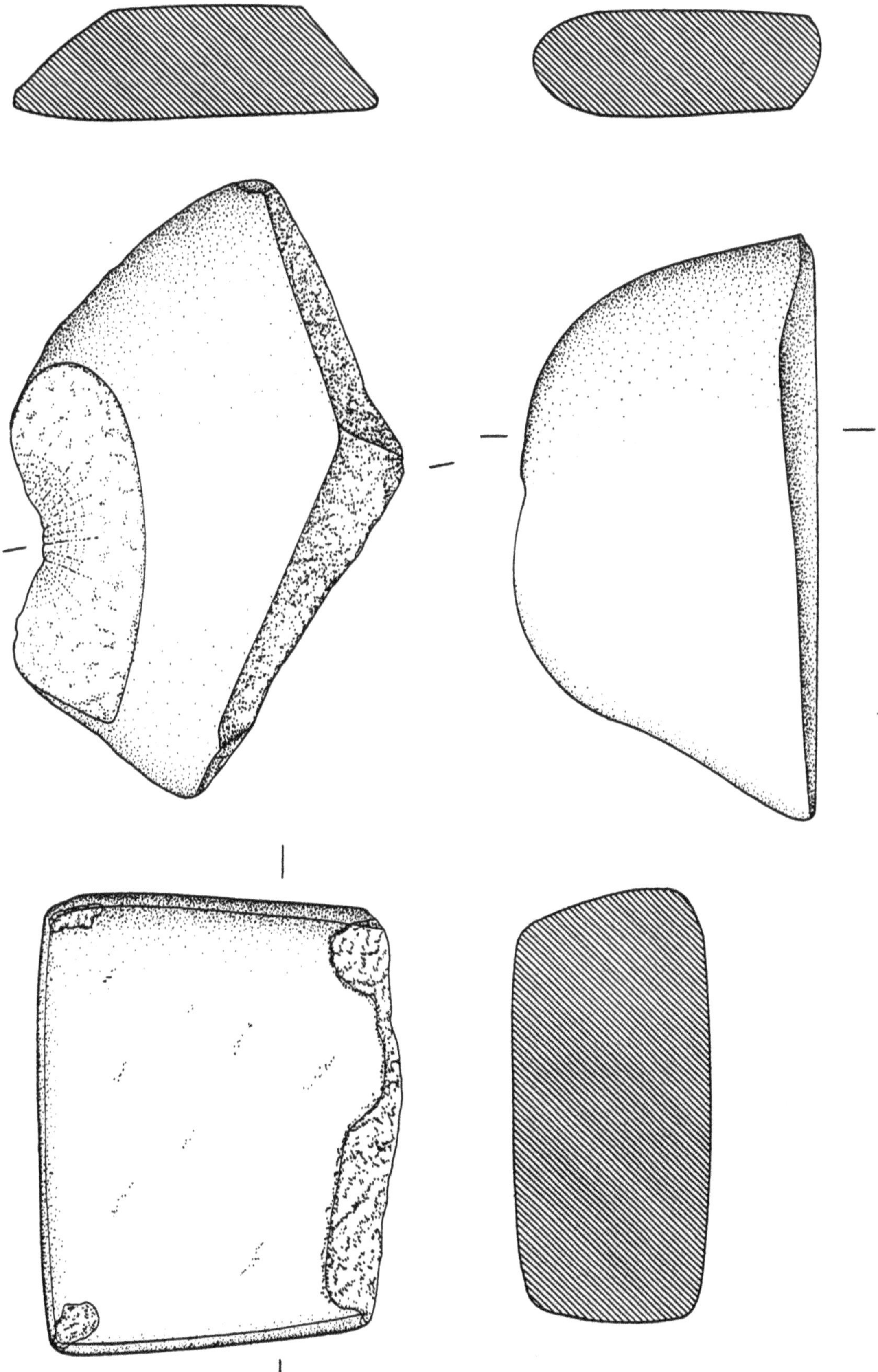

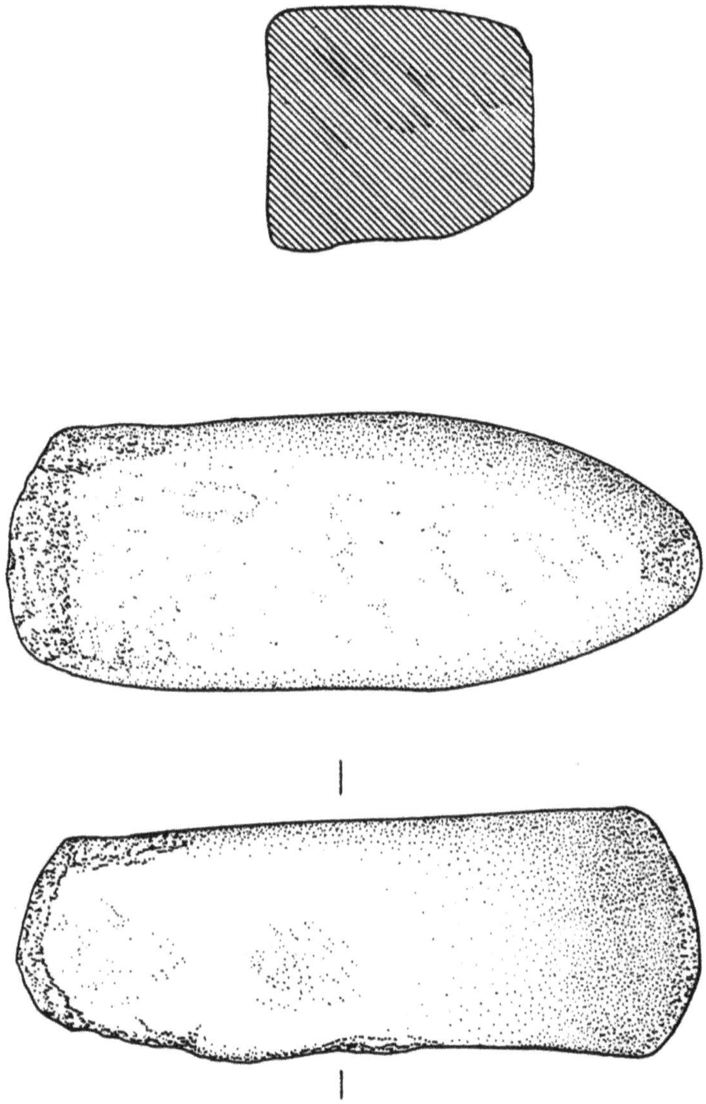

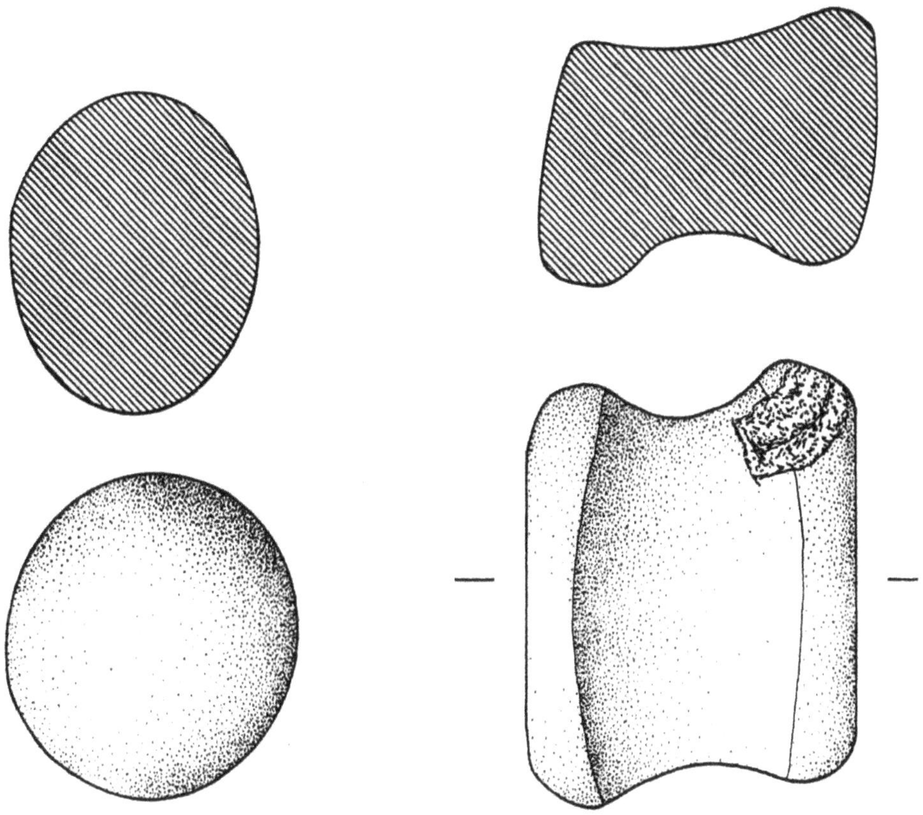

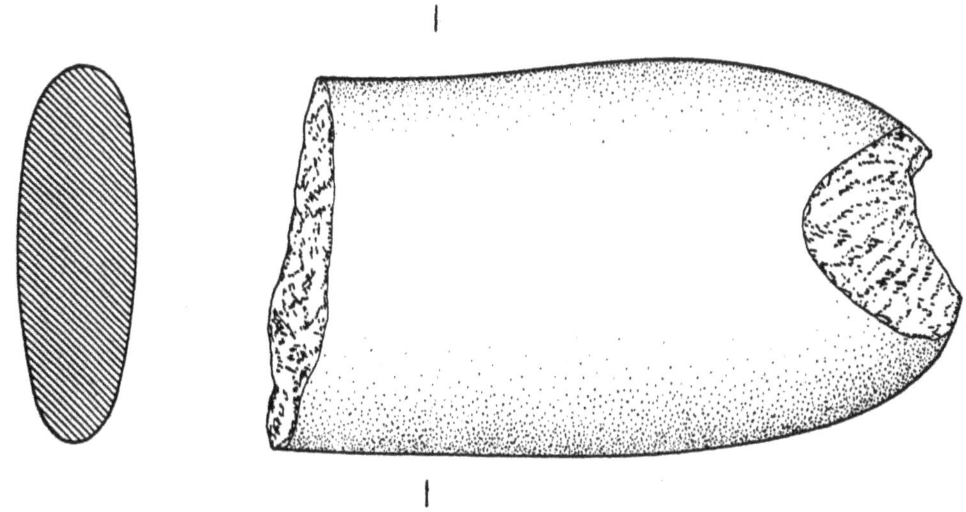

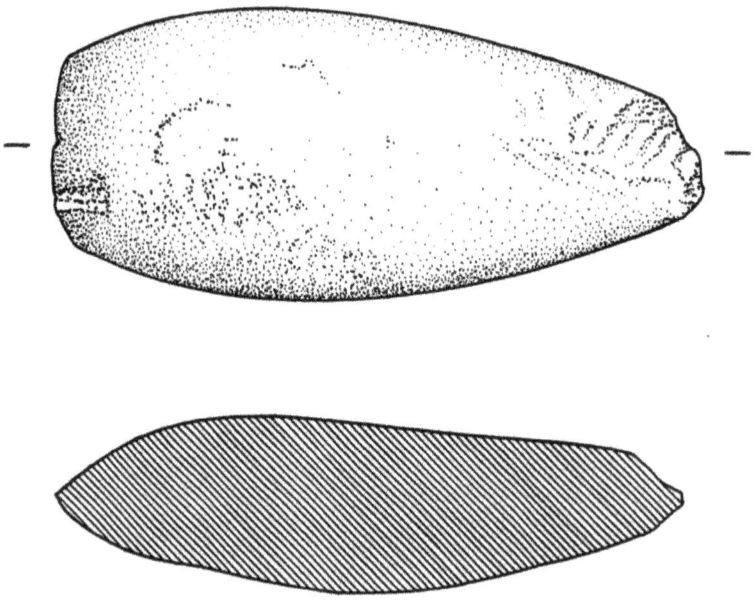

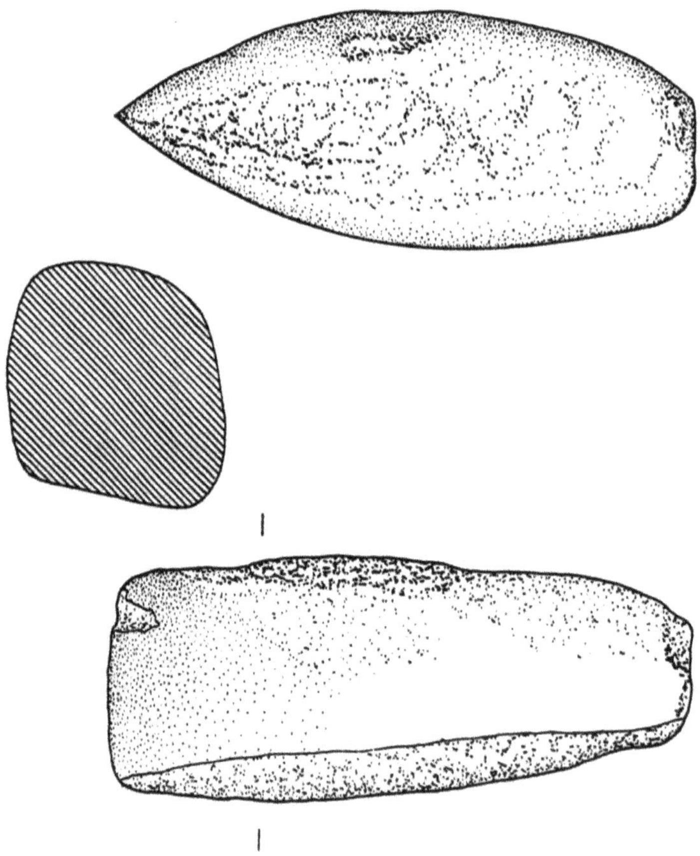

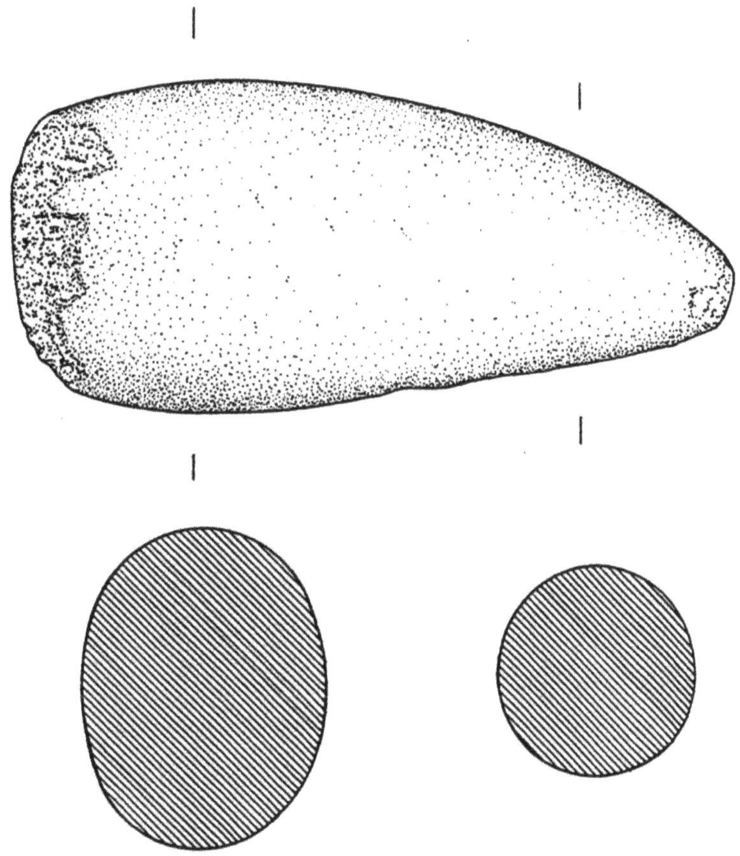

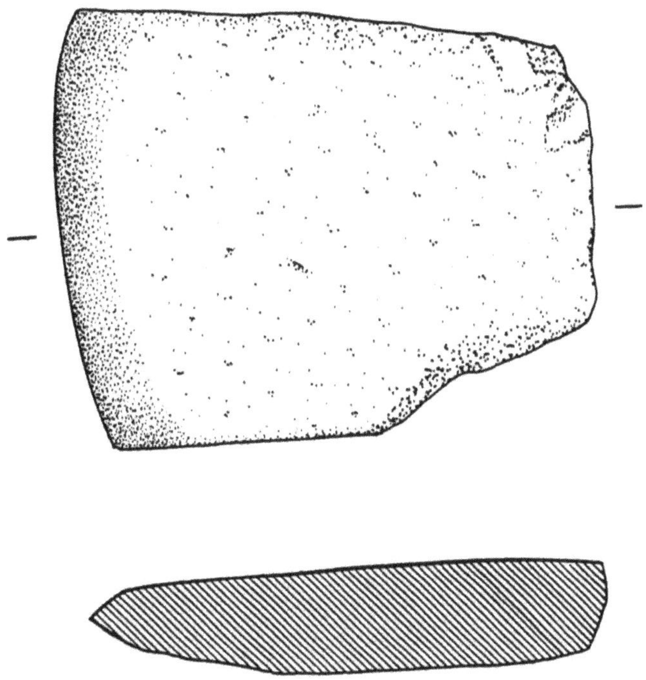

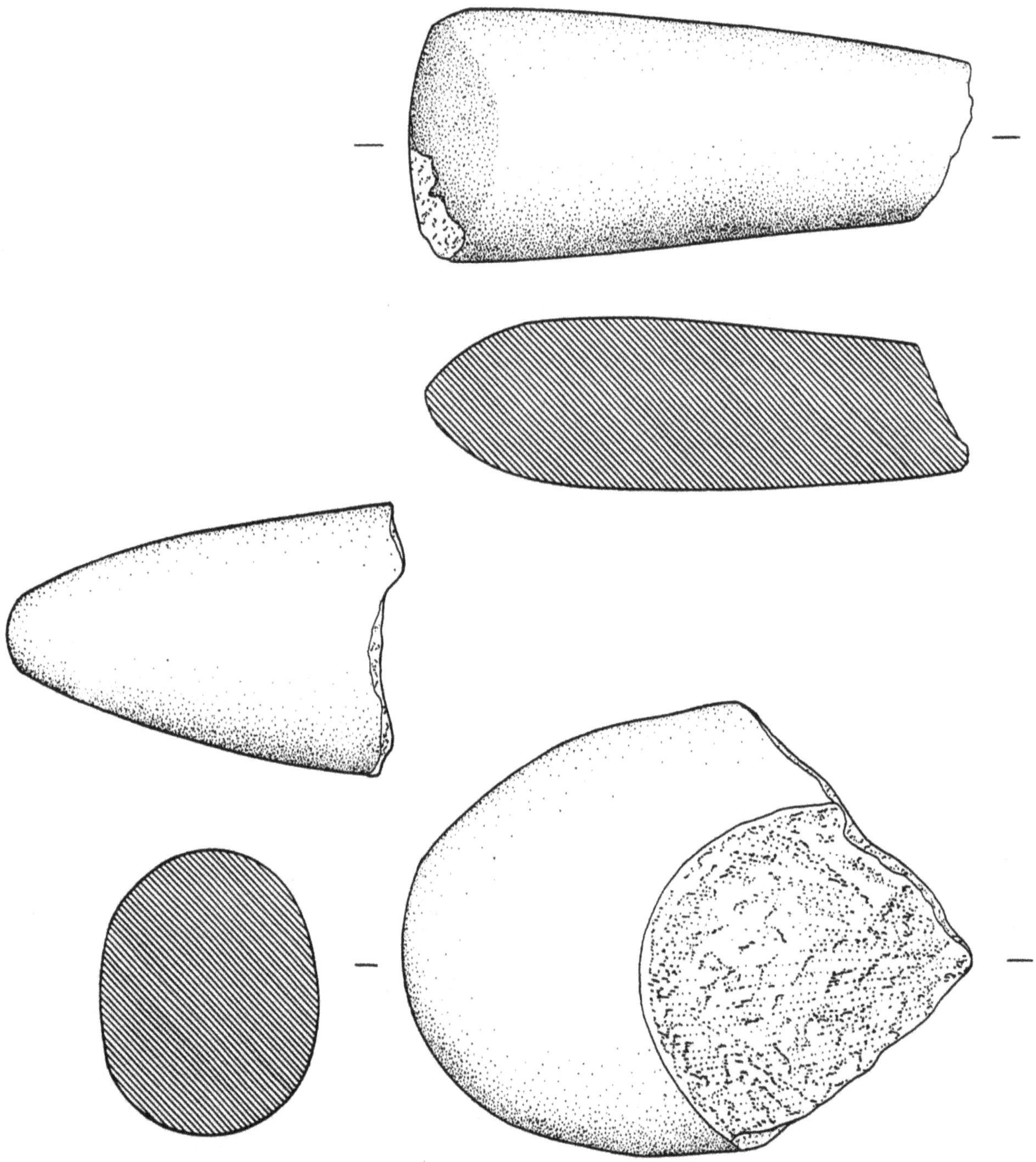

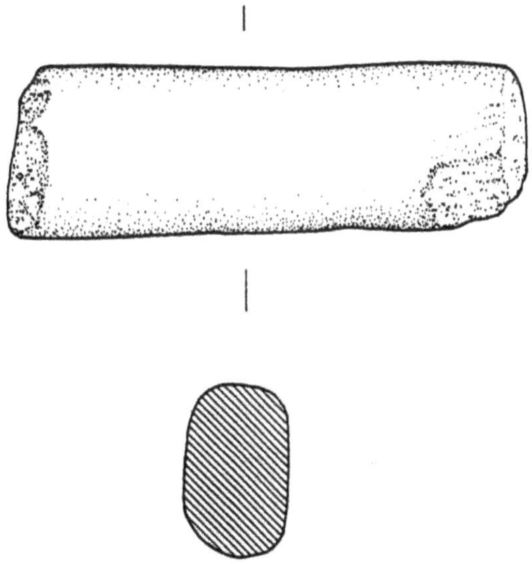

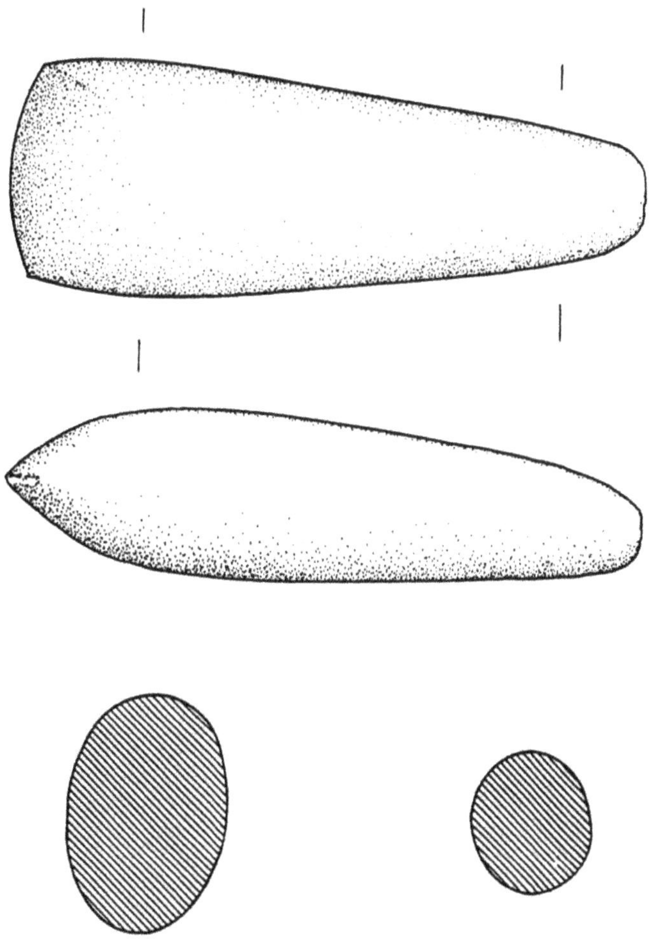

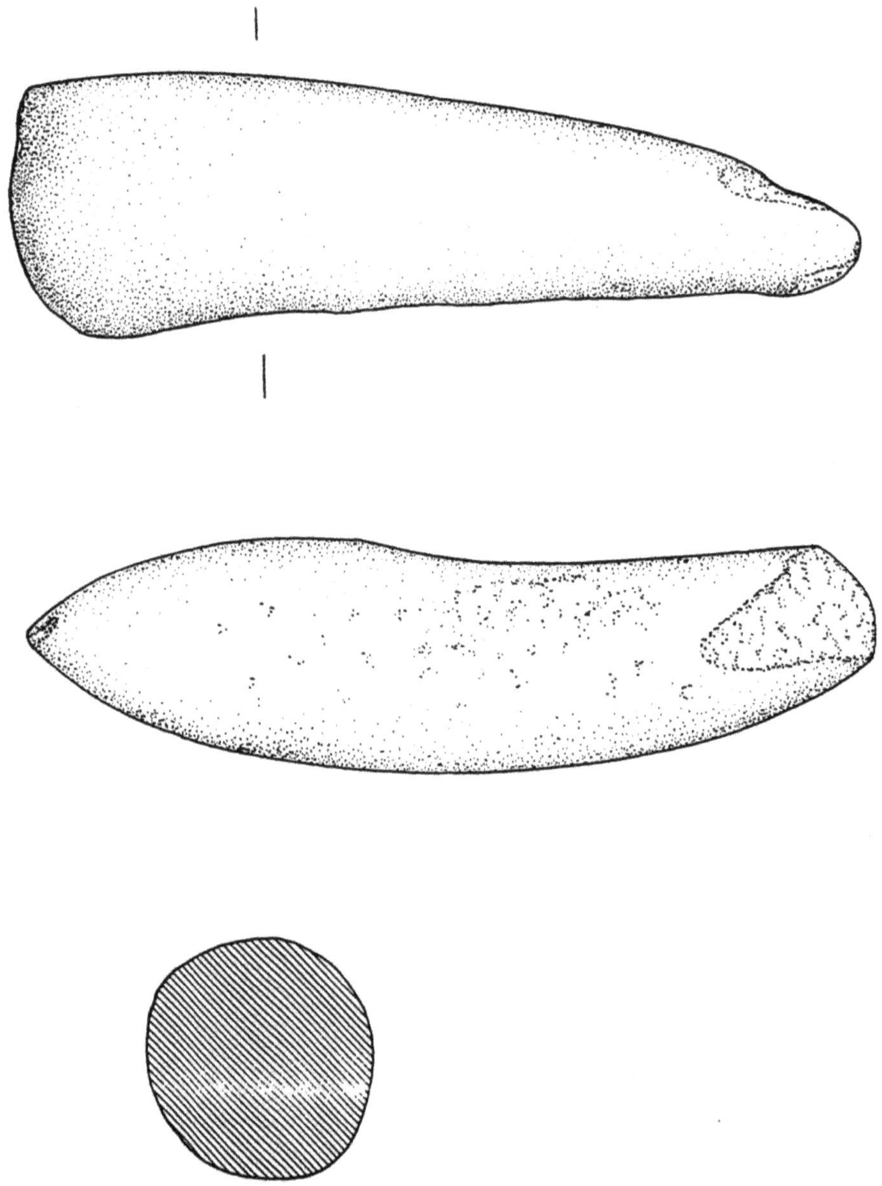

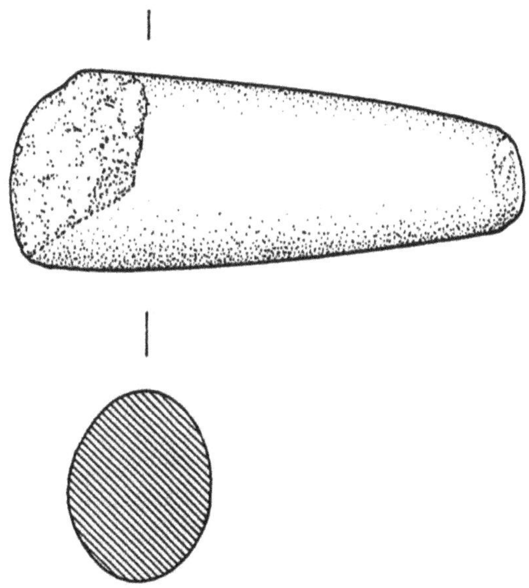

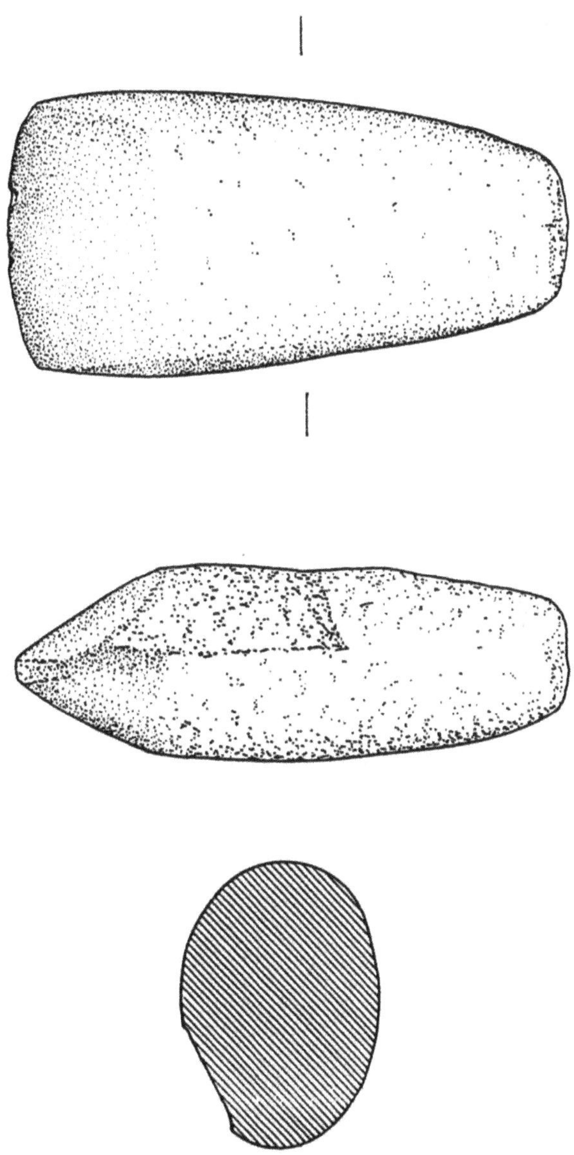

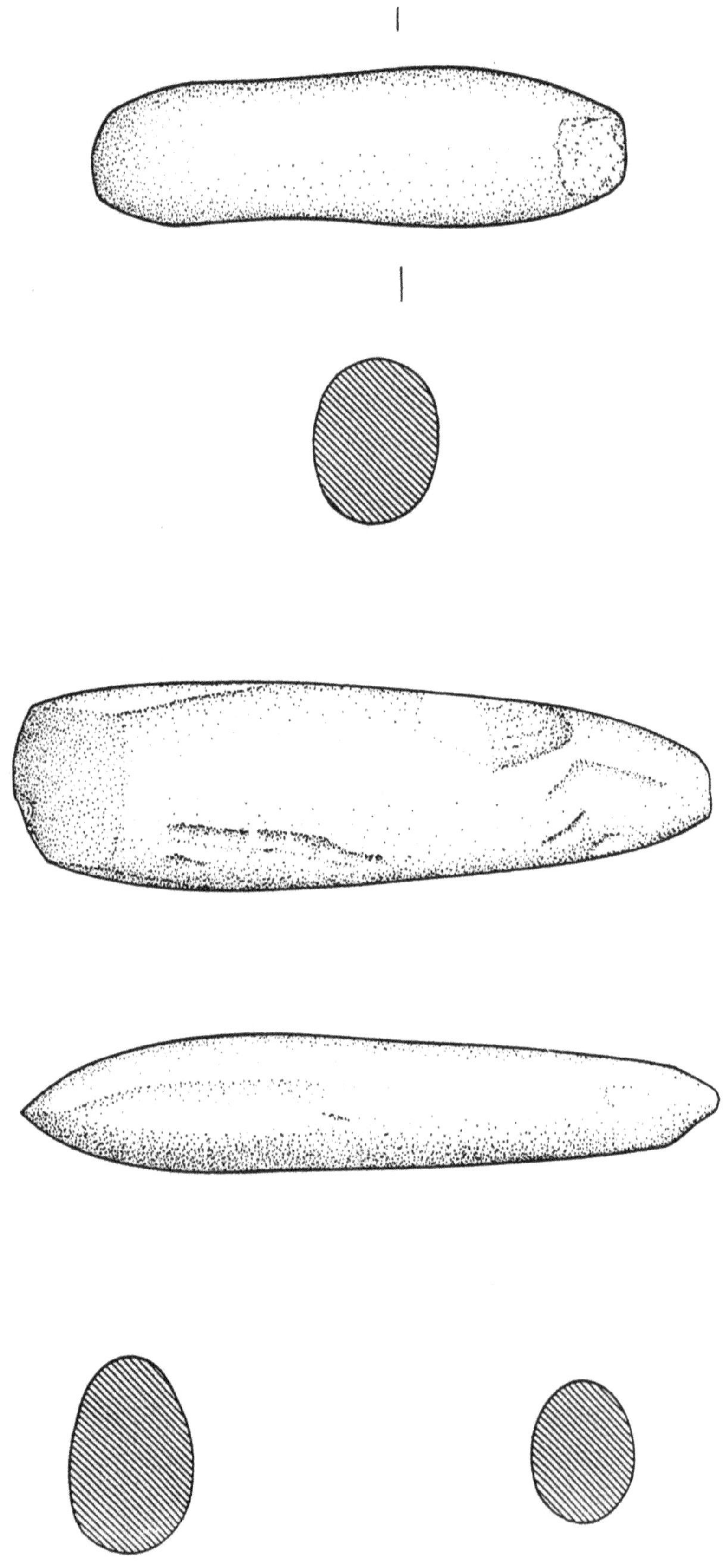

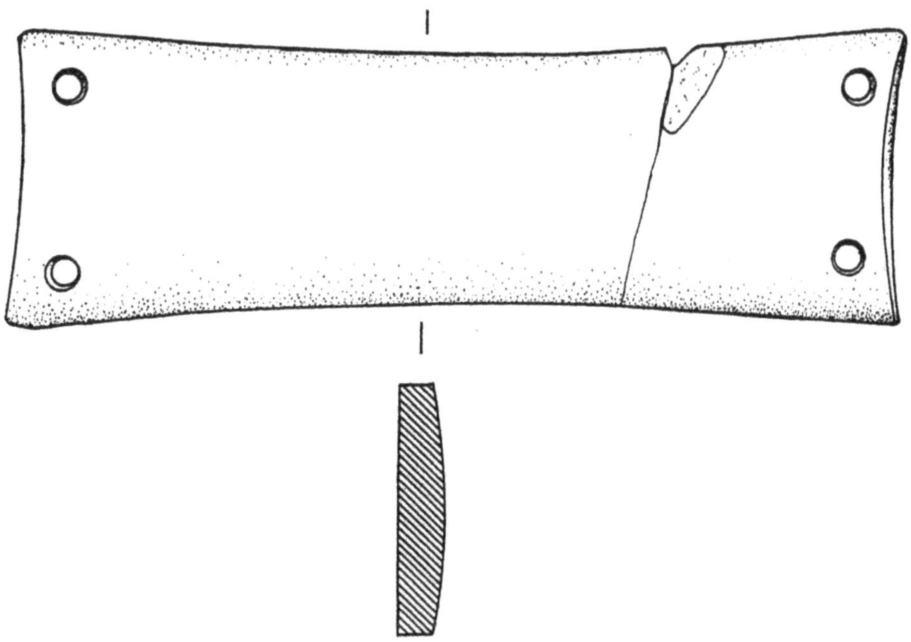

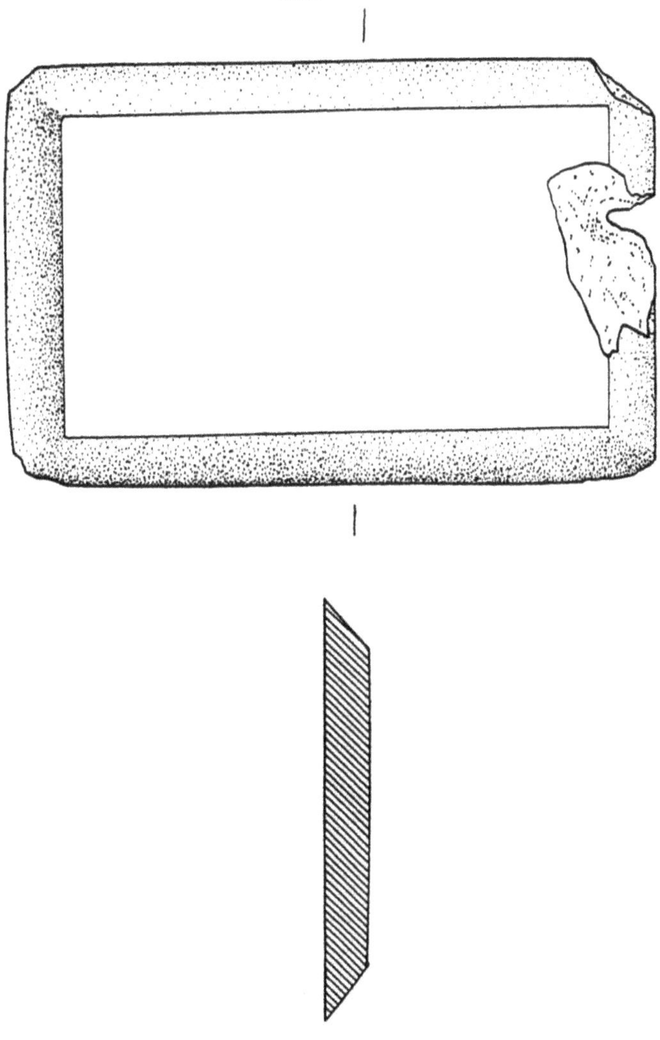

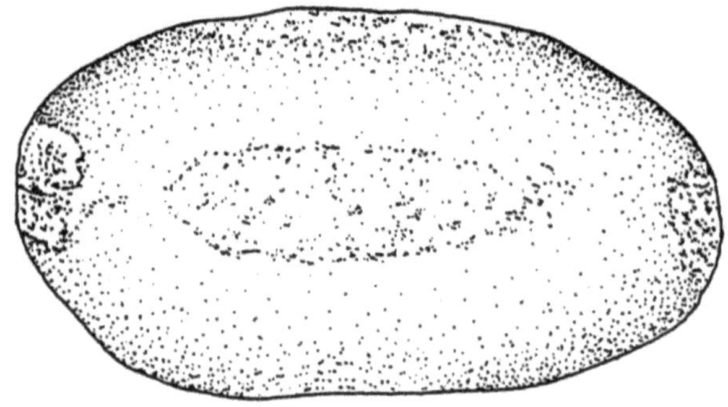

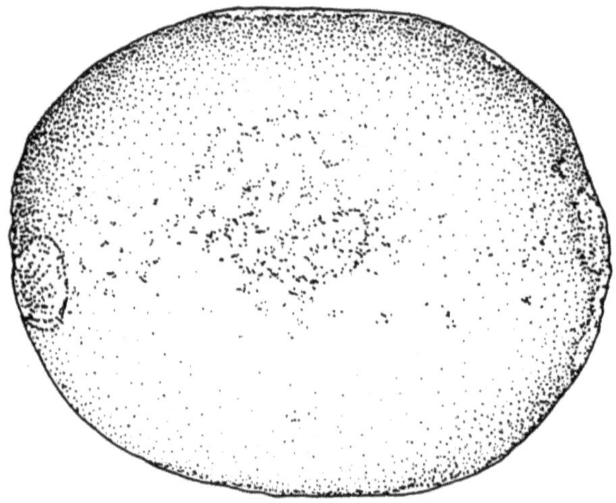

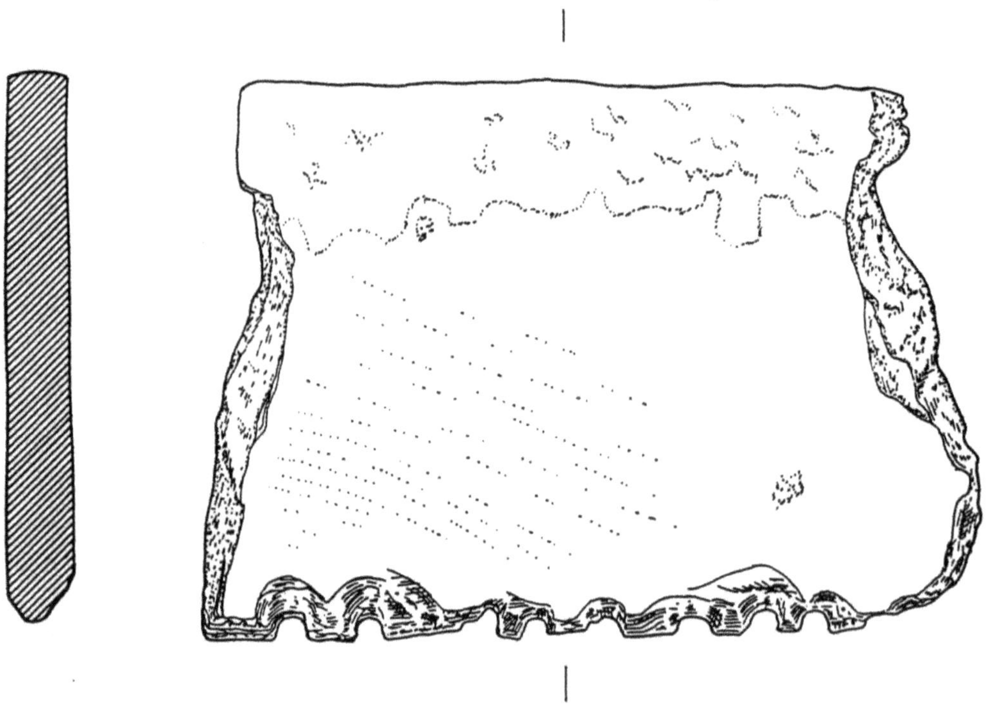

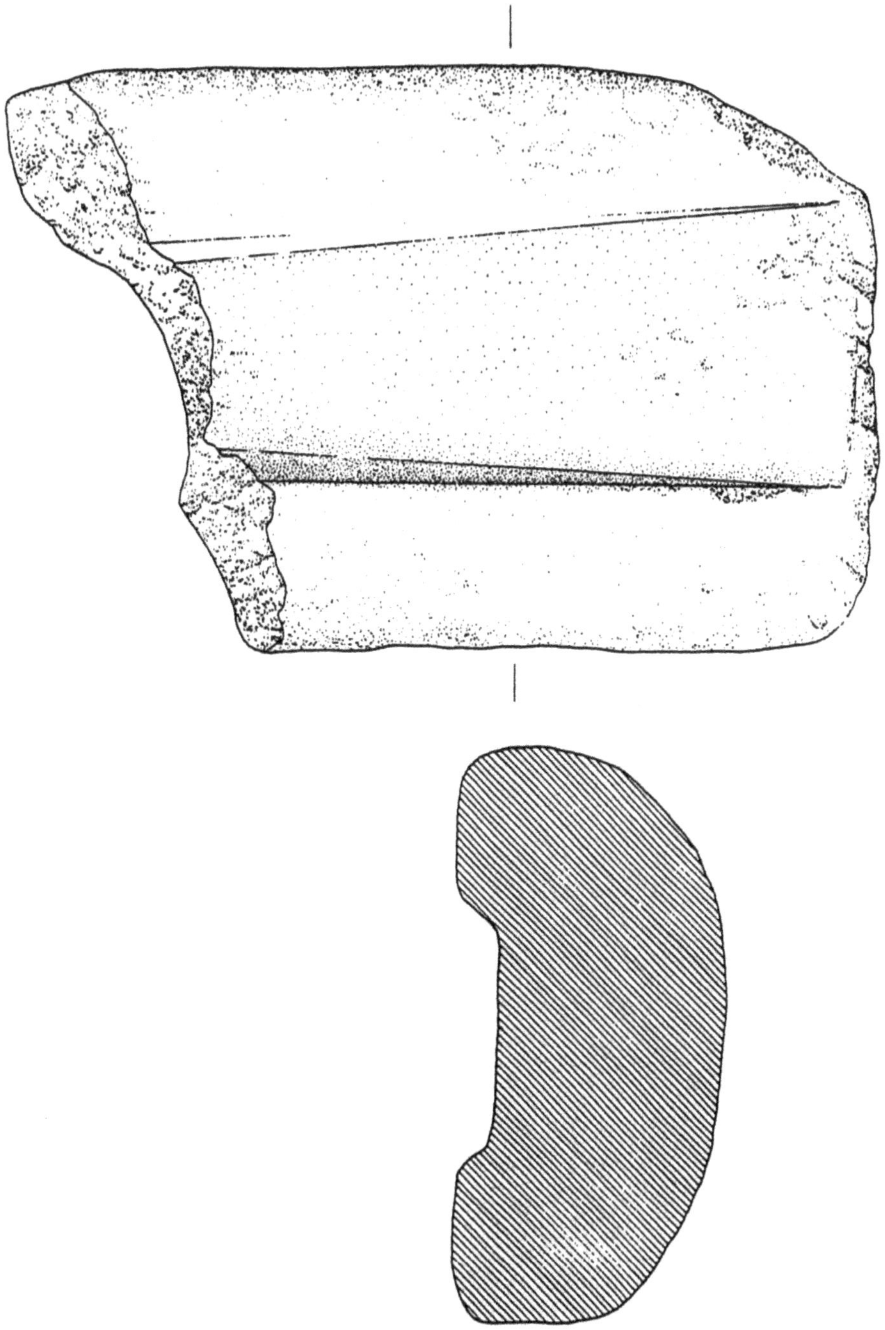

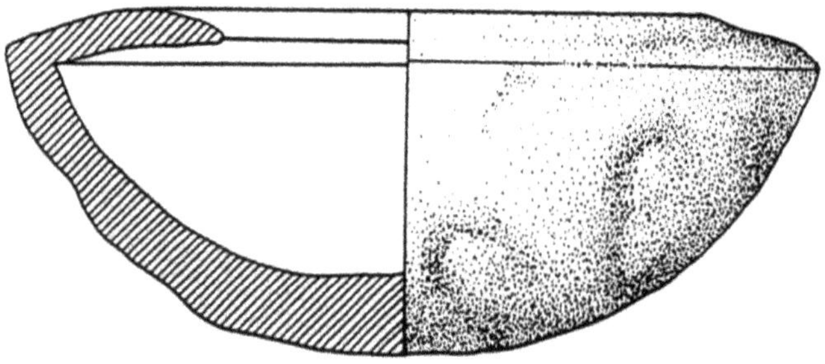

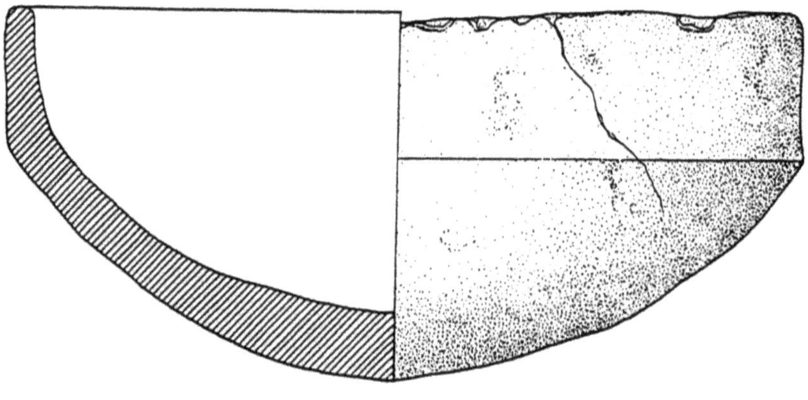

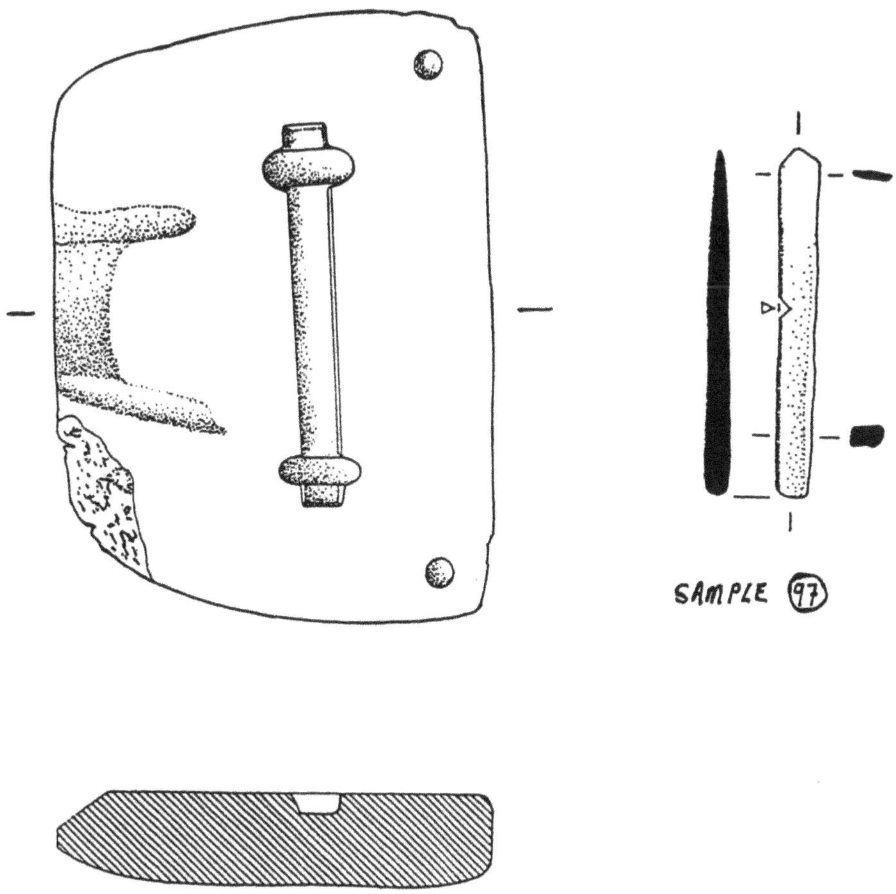
SAMPLE 97

APPENDIX II

LABORATORY ANALYSIS and
DRAWINGS OF METAL ARTEFACTS
Museo Nacional de Arte Romano, Merdia

from LOGROSAN
El Cerro de San Cristobal

collected by SOS BAYNAT 1950-1960

drawn by JOSEFINA MOLINA GARCIA
1994-1995

sampled and analysed by CRAIG MERIDETH
1995

Description and Analysis of Metal Artefacts Logrosan

The ten cut samples were taken from the Sos Baynat collected Logrosan artefacts stored at the Museo Nacional De Arte Romano, Merdia, Extremadura, Espana (Spain). I wish to thank
the Secretary of the Museum Augustin Velezquez Jimenez for his help in making it possible for me to sample the Logrosan
metal artefacts.

All cut samples were mounted in resin and finely polished and carbon coated before being put into the SEM specimen chamber.

Scanning Electron Microscopy using Energy-Dispersive X-ray as a qualitative technique was used to analysis theelements present in the ten metal samples. The less than 100% analytical results represent any oxygen or carbon that was present in the samples and which the SEM/EDS analytical method doe not detect. A minimum of three areas were analysed on each sample to ensure that the initial analysis
was correct.

The SEM/EDS analytical work was done in the Wolfsan Laboratories, Institute of Archaeology, University College London.

Choosing the area in which to take the sample and the size of the sample to be cut was dictated by having to choose areas which were not heavily corroded and which would not in any way distort the original appearance of the artefact.

Another important consideration in choosing the sample area was to try and choose a location which would also offer a acceptable cross section for a metallography-microstructure examination study.

When cutting the ten Logroasan samples I weighed the total weight of the artefacts and also in some cases if possible the weight of the cut samples.

Sample No 110 A Bronze Flat Axe;
total sample weight 690.9 grams
cut sample weight 04.5 grams

The sample was cut from one of the edges of the flat axe as I felt that both the butt and splayed blade of the axe had been heavily damaged by secondary use or misuse and might not give a true representative sample.

A total of nine areas were analysed in the cut and mounted sample. The style of the flat axe should represent a Early to Middle Bronze age artefact (Savory, 1980, 44); but the presence of 1.38%-12.51% Lead (Pb) in all of the areas analysed attribute the Flat Axe probably being of Late Bronze Age origin (Tylecote, 1962, 43-45). The lead would probably have been added to help produce a better running casting (Tylecote, 1992, 35).

Three areas analysed (Area 3, Area 4, and Area 9) probably represent areas of corrosion which accounts for the large amounts of Tin (Sn) in Area 4 and Area 9 and the large amount of Lead (Pb) in Area 3. In these three areas the copper has corroded away and left enriched deposits of Tin and Lead (Cronyn, 1990, 166, 218).

Sample No 111 A cast Bronze (?) Dolphin (?)
shaped object;
total sample weight 138.5 grams

The uncommon shape of Sample No 111 possibly represents a cast cult object ?

A total of three areas were analysed in the cut and mounted sample. The analysis on all three areas was fairly consistent. This sample analysed as a very well produced bronze (Cu-Sn) mixture.

Sample No 112 A cast Bronze (?) Monkey/Dog (?)
total sample weight 26.7 grams

The uncommon shape of Sample No 112 which was possibly attached to a wooden rod, also possibly represents a cast cult object.

A total of three areas were analysed in the cut and mounted sample. The analysis of Area 1 and Area 2 were very consistent and represent a almost ideal Cu-Sn bronze mixture. Area 3 with its high content of 13.48% Sn might represent a corroded Tin enriched area of the bronze sample.

Sample No 113 An arsenical/copper Beaker Palmela Point
total sample weight 5.8 grams

All three areas analysed were very uniform in the results. The arsenic present only varied from 1.78% to 2.45%. The analytical results of Sample No 113 duplicate other analysed Beaker Period Palmela Points (Oddy, 1991, 52-62).

Sample No 114 A cast (?) (Iberian/Roman) (?)
Shaped Small Figure
total sample weight 17.5 grams

Sample no 114 was very corroded on all surface areas. Three areas were analysed and all three results varied considerably.

The lead (Pb) 20.38% and Tin (Sn) 6.69% present in Area 1 most probably represents the norm for the sample No 114. Area 2 probably represents a corroded area which has given rise to the high tin (SN) 12.81% content being present. Area 3 is a lead (Pb) 88.15% rich area which has been analysed. The lead enrichment areas are a result of the insolubility effect of smelting lead with copper (Scott, 1991, 23).

The Copper (Cu), Tin (Sn) and Lead (Pb) mixture of Sample No 114 point to a Late Bronze Age or Roman origin.

Sample No 115 A Bronze cast
(?) Iberian/Roman (?) object
total sample weight 17.7 grams

Three areas were analysed on this sample. Area 1 with its high percentage of Tin (Sn) 30.39% has to represent a corroded area which is Tin enriched due to Copper loss during the corrosion process (Cronyn, 1990, 218).

Area 3 and Area 4 which are very similar in analytical results represent the true elemental makeup of the artefact.
The high Tin content would tend to indicate that the cast object was meant to be highly polished and to present a high lustre surface in its finished phase (Scott, 1991, 25-27)

Sample No 116 A Bronze Arrow Point- origin unknown
sample total weight 04.9 grams

Five areas were analysed on this cut sample. Area 1 and Area 2 analytical results were either too high over 100% or to low under 65% and therefore rejected.

Area 3, Area 4 and Area 5 are uniform in their Copper (Cu) content and vary slightly in the Tin (Sn) content 4.29 %, 6.11% and 8.56%.

Sample no 116 is a very well produced bronze Arrow point. The archaeology period it represents is not known at present. It could be Late Bronze Age or Roman from the percentages of the Cu-Sn mixture present.

Sample No 117 A small Bronze (?) awl/needle (?)
total sample weight 01.8 grams
cut sample weight 0.2 grams

Six areas were analysed on this sample The analytical results from Areas 1, 2, and 3 were rejected as they were in excess of 100%. Area 4, Area 5 and Area 6 were very uniform in the analytical results with copper content varying only 2.36% and tin content varying only 0.55%.

It is not exactly clear/obvious if Sample No 117 is a purpose made artefact or only a casting waste droplet.

Sample No 118 A very small Bronze (?) droplet (?)
total sample weight 02.2 grams
cut sample weight 0.1 grams

Three areas were analysed on this very small cut sample. The variation in all three results was negligible and represents a good bronze mixture.

Sample No 118 must surely be a waste droplet from a refining or casting process.

Sample No 119 A small Copper (?) Roman (?)
Ornament/with Zn, Sn, Pb, Fe present
total sample weight 01.7 grams

Four areas were analysed on Sample No 119. The analytical results from Area 3 were rejected as they were well in excess of 100% total.

Sample No 119 was different from any of the other nine samples from Logrosan in that it contained Copper (Cu), Tin (Sn), Lead (Pb), small traces of Iron (Fe) and Zinc (Zn). The tin content was not high 1.72%-3.19%. The Lead content was also not high 1.75% - 3.91%. The traces of Iron present in all three areas being under 0.50%. The Zinc content was 8.08% - 8.77% in the three areas analysed.

It has been very rare to find Zinc alloyed with Copper before the Roman period It was in the Roman period that Zinc was widely used and alloyed with Copper Lead and Tin in decorative jewellery and figurative objects (Waldbaum,1983, 174), (Tylecote, 1962, 53-56)

There remains the possibility that Sample No 119 with its Copper, Zinc, Lead, Tin and traces of Iron might be Medieval in origin (Tylecote, 1962, 57).

EL CERRO DE		%	%	%	%	%	%	%	%	%	%	%	%
SAN CRISTOBAL		Cu	Sn	As	Pb	S	Fe	Si	Ca	Al	Zn	Na	TOTAL
LOGROSAN													
Museo Nacional													
de Arte Romano													
Merdia/Badajoz													
Sample No 110	Area 1	72.69	10.53		3.26	0.41							86.91
Axe	Area 2	73.78	9.44		2.86	0.18							86.47
4.12.95	Area 3	23.49	4.48		12.51	2.40						4.29	47.18
	Area 4	21.86	36.13		7.99	4.17							66.94
	Area 5	58.92	5.41		1.38	13.12							78.84
	Area 6	82.44	12.73		1.82	0.31							97.30
	Area 7	82.06	14.62		1.55	0.23							98.47
	Area 8	77.13	14.38		3.84	0.65							96.01
	Area 9	47.43	22.66		9.06	0.94							80.10
Sample No 111	Area 1	83.58	10.13										93.72
Dolphin Shape(?)	Area 2	79.65	8.95			4.39							93.00
1.12.95	Area 3	85.37	6.61										91.99
SampleNo 112	Area 1	91.55	7.23										98.79
Monkey/Dog (?)	Area 2	95.19	7.33										102.53
1.12.95	Area 3	88.61	13.48										102.09
Sample No 113	Area 1	91.58		2.29									93.82
Palmela Point	Area 2	92.50		1.78									94.29
1.12.95	Area 3	92.7		2.45									95.15
Sample No 114	Area 1	65.31	6.69		20.38								92.39
Iberian/Roman (?)	Area 2	83.25	12.81		1.69								97.46
1.12.95	Area 3	0.56			88.15								89.15
Sample No 115	Area 1	69.71	30.39										100.16
Roman (?)	Area 3	87.75	13.92										101.69
1.12.95	Area 4	86.41	15.39										101.86
Sample No 116	Area 3	89.96	4.29			0.10							94.37
Arrow Point	Area 4	87.78	6.11										93.94
1.12.95	Area 5	86.11	8.56										94.67
Sample No 117	Area 4	83.22	8.64										91.87
Small Awl	Area 5	81.65	9.35										91.03
1.12.95	Area 6	84.01	9.19			0.38							93.59
Sample No 118	Area 1	81.58	11.68										93.26
Small Metal Droplet	Area 2	87.06	11.27										98.33
29.11.95	Area 3	81.52	11.34										92.86
Sample No 119	Area 1	85.73	3.52		3.82		0.48				8.77		102.34
Small Ornament	Area 2	84.7	1.72		1.75		0.44				8.31		96.94
Medieval (?)	Area 4	81.49	3.19		3.91		0.42				8.08		97.11
1.12.95													

EL CERRO DE		%	%	%	%	%	%	%	%	%	%	%	%
SAN CRISTOBAL		Cu	Sn	As	Pb	S	Fe	Si	Ca	Al	Zn	Na	TOTAL
LOGROSAN													
Visit 4													
May-95													
METAL	Area 1	86.86	9.31										96.18
Sample No 109	Area 2	84.51	10.11										94.62
Small Fragment	Area 3	82.29	9.73			0.25							92.28
29.11.95	Area 4	81.07	9.96			0.23							91.27

Sample No 109 A very small copper fragment.

This fragment was found on the surface 1.5.1995, approximately fifty meters directly West of the 677 geological survey post.

Due to its small seize the complete sample was mounted. Four areas were analysed on Sample No 109. All four areas represent a perfect bronze metal mix consisting of 9%-10% tin and 81%-87% copper. This sample must represent a fragment from a bronze casting process.

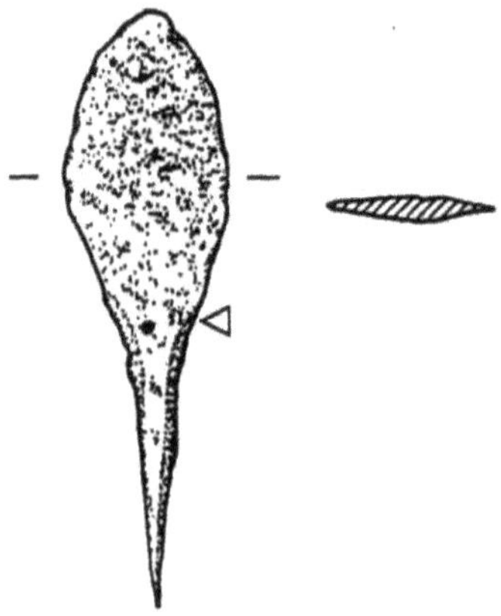

Sample No 113

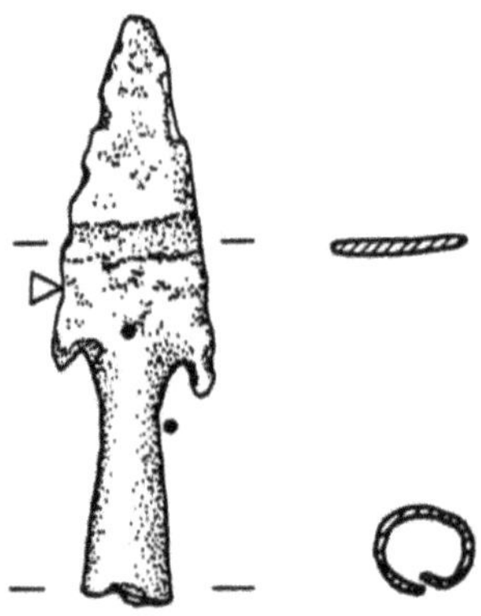

Sample No 116

◁ cut sample

• drilled sample

○ drilled sample
taken from reverse side

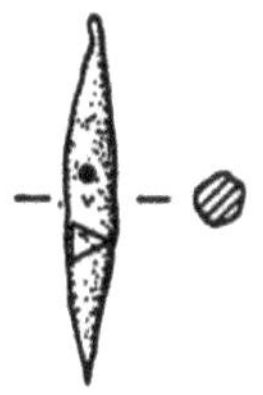

Sample No 117

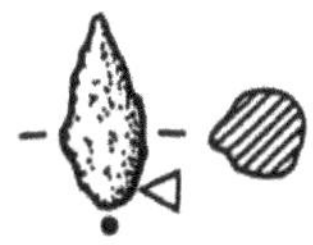

Sample No 118

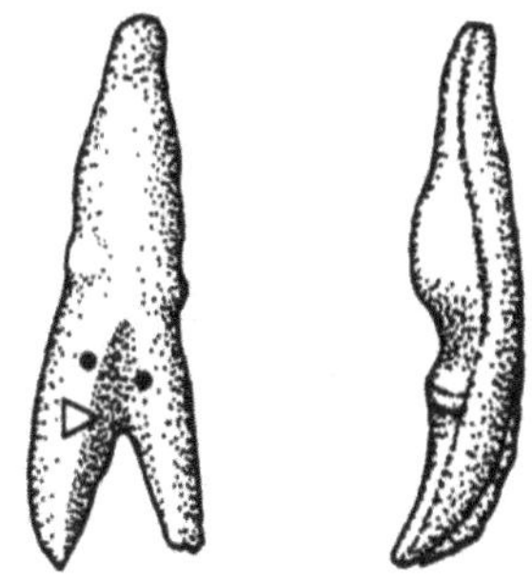

Sample No 114

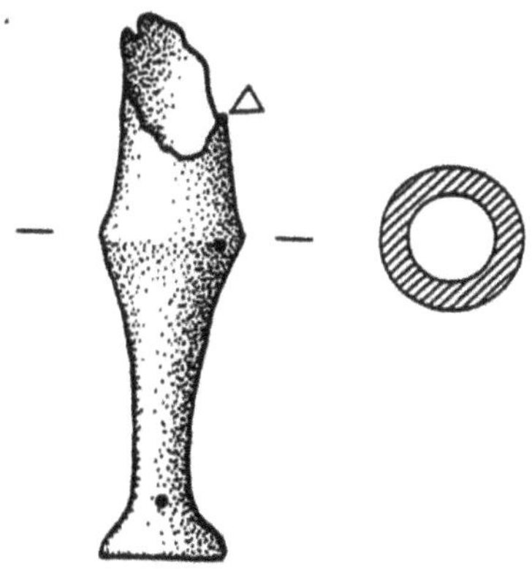

Sample No 115

Sample No 119

◁ cut sample

● drilled sample

○ drilled sample
taken from reverse side

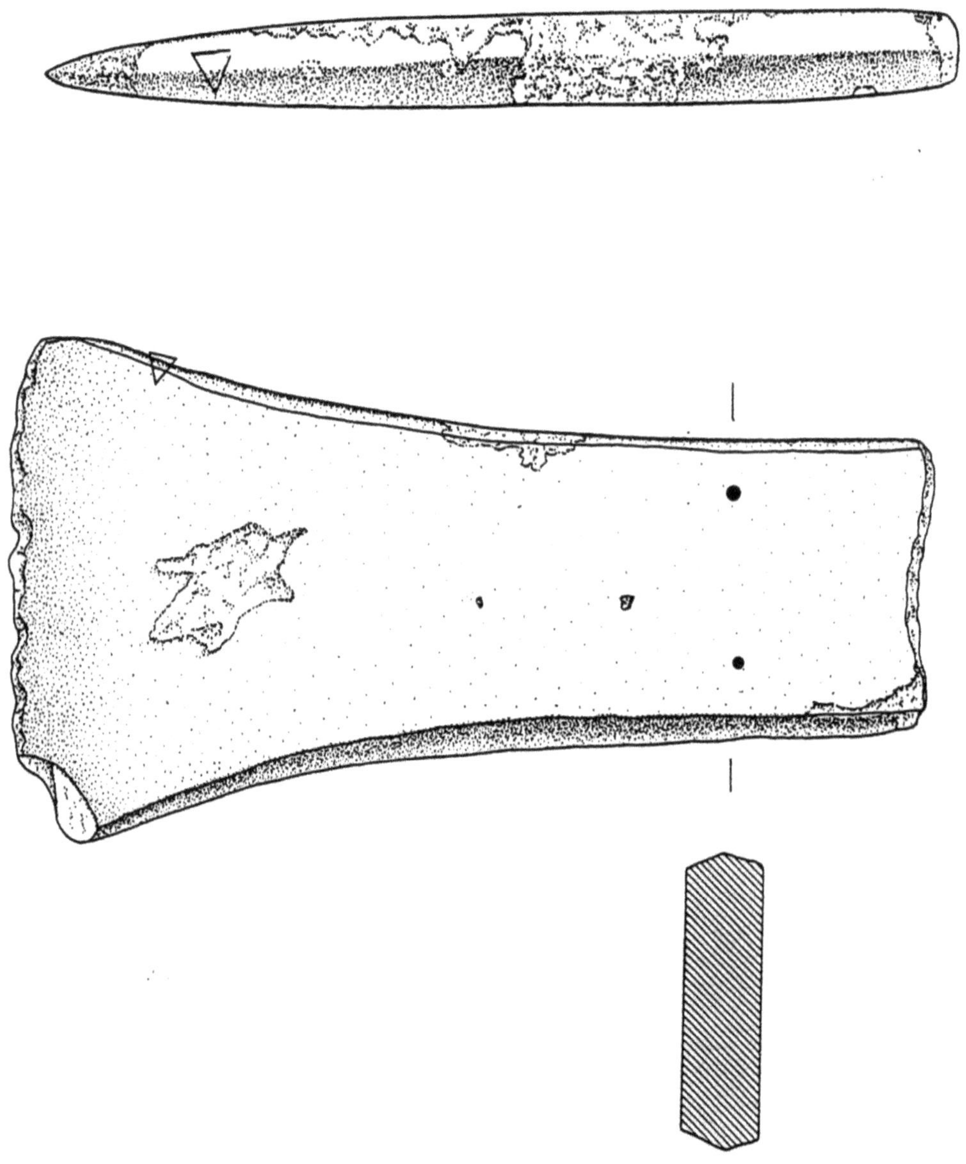

Sample No 110

◁ cut sample

• drilled sample

○ drilled sample
taken from reverse side

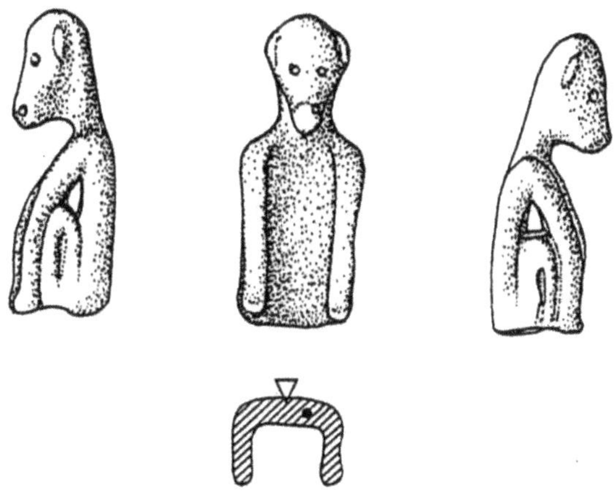

Sample No 112

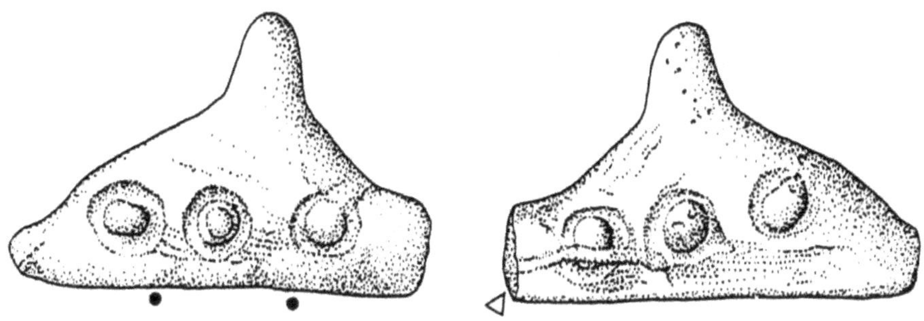

Sample No 111

◁ cut sample

• drilled sample

○ drilled sample
taken from reverse side

www.ingramcontent.com/pod-product-compliance
Lightning Source LLC
LaVergne TN
LVHW071208210926
848482LV00015B/750
9780860548928